Geometry and Fractions with Geoboards

Problem-Solving Activities
Grades 3-6

by
Barbara Bando Irvin, Ph.D.

Edited by: Jerome Pohlen
Illustrations by: Denise Gardener
Design by: Ellen Hart/Concepts Plus, Inc.

ISBN: 1-56911-939-2

Printed in the United States of America.

Table of Contents

Introduction

Geometry and Fractions with Geoboards contains 38 hands-on activities to help students in grades 3-6 learn and reinforce basic concepts and skills associated with plane geometry, measurement, and fractions. The activities also provide students with the opportunity to enhance spatial skills and apply problem-solving strategies.

The activities are designed to be used with a double-sided geoboard having a 5 x 5 square array (25 pins) on one side and a 12-pin circular pattern on the other side. Recording sheets containing nine smaller geoboards of each type are provided on pages 63 and 64. The black-line masters may be duplicated to use in the classroom, to make presentations, or to show solutions on the overhead projector.

A Glossary of the vocabulary terms used in the activities is on page 62. You may wish to duplicate copies for your students. Selected Solutions are provided at the end of the book (pages 54-61).

Overview of Content

The activities are presented in two major sections by geoboard type to help you integrate geometry and fraction lessons into your mathematics curriculum. Activities on pages 7–34 utilize the 5 x 5-pin geoboard; activities on pages 35–53 are geared for the circular geoboard.

NCTM's *Curriculum and Evaluation Standards for School Mathematics* (1989) was used as a guideline to develop activities in this book. Specific goals for geometry, measurement, and fractions are:

- to identify, describe, and classify geometric shapes
- to represent and solve problems using geometric models
- to explore transformations of geometric figures
- to understand and apply geometric properties and relationships
- to relate geometric ideas to number ideas
- to understand the attributes of length, area, and angle
- to develop concepts related to units of measure
- to model and develop concepts of fractions
- to find equivalent fractions
- to model addition of fractions

The Table of Contents lists the concept(s) or skill(s) explored in each activity. These should act as a springboards for more explorations using a 5 x 5-pin geoboard or a circular geoboard. Modify the activities to create games, riddles, writing assignments, and assessment tasks.

Using the Activities

Each section of *Geometry and Fractions with Geoboards* begins with Teaching Notes to provide a list of vocabulary terms, mathematical explanations, and suggestions for classroom use. Vocabulary definitions are provided in the Glossary on page 62.

Although many geometry and fraction concepts are developed separately, sometimes they are explored together in an activity. For example, area and fractional parts of regions are compatible, as are central angles and fractions. Many of the activities are open-ended and have multiple solutions. Extend each activity by urging students to find all possible relationships and answers. Have extra copies of the recording sheets (pages 63-64) available for them to record additional solutions.

Although the rubber bands go *around* the pins on the geoboards, tell students to draw line segments from point to point on the lesson pages and recording sheets. Ask them to use a straightedge or ruler to draw the line segments. Permit them to use calculators to solve computation problems. If they wish to verify an angle measurement, they may use protractors.

Basic Concepts and Problem Solving

The geoboard is a hands-on tool that embraces a variety of mathematical concepts, relationships, and thought-provoking situations. Definitions and properties of geometric figures can be used to describe a shape or find the measure of its angles. Techniques for solving problems can help students set up and analyze mathematical situations. Problem-solving strategies used in this book include:

- Using a model
- Using visual reasoning
- Working backward
- Drawing a picture
- Guessing and checking
- Using logical reasoning
- Looking for patterns
- Making an organized list

Concepts and Vocabulary

Knowing the correct mathematical terminology can help students communicate their ideas to one another. Have children record important vocabulary terms in their journals with a definition, sketch, and example. Students also can draw pictures of geometric figures using copies of pages 63-64.

Cooperative Learning and Communication

Students should be encouraged to work together in cooperative pairs or small groups to make a discovery, do an activity, or find all possible solutions. Encourage students to verify and interpret results by having them talk or write about the strategies and procedures they used to solve a problem. Have students incorporate the activities' vocabulary terms in discussions and written assignments.

Extensions and Creativity

Many of the activities may be extended by asking "What if?" questions. Capitalize on students who present alternate approaches to solving problems, which can help to expand other students' thinking. Urge children to create their own problems and puzzles to challenge each other.

Using the Geoboards

Before students begin, establish class rules about using the geoboards and rubber bands. Tell students that rubber bands can be harmful if not properly used. Make sure they understand that the rules are for their safety, and everyone must follow the rules in order to work with the materials.

Free Exploration

Many students will have had previous opportunities to work with a 5 x 5 geoboard in the primary grades. Have them show and talk about different line segments and geometric shapes. Students familiar with a 5 x 5 geoboard should do some exploration activities on the circular geoboard. Observe and listen to students as they show circles, angles, and shapes. Ask them to share what they have discovered with the rest of the class.

About the 5 x 5 Geoboard

The 5 x 5 square array of pins can be used of show line segments, angles, and shapes. Use a nondiagonal line segment between two adjacent pins to indicate ***1 unit*** of length to find the length of a line segments or the perimeter of a shape. Use a square composed of connecting four adjacent points with nondiagonal lines to be used as ***1 square unit*** to find the area of a shape. Diagonal line segments require estimation to find their lengths.

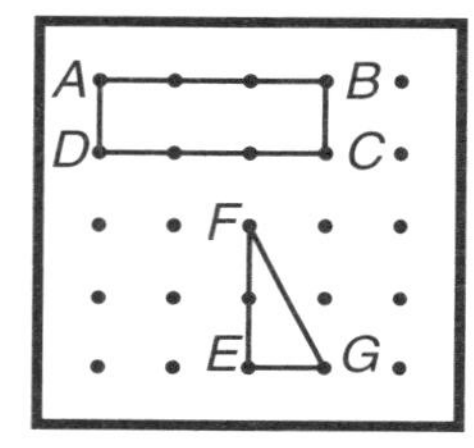

1 unit of ***length***
Line segment $\overline{AB}$ is 3 units long.
Perimeter of rectangle ***ABCD*** is 8 units.
Line segment $\overline{FG}$ is about 2¼ units long.
Perimeter of triangle ***EFG*** is about 5¼ units.

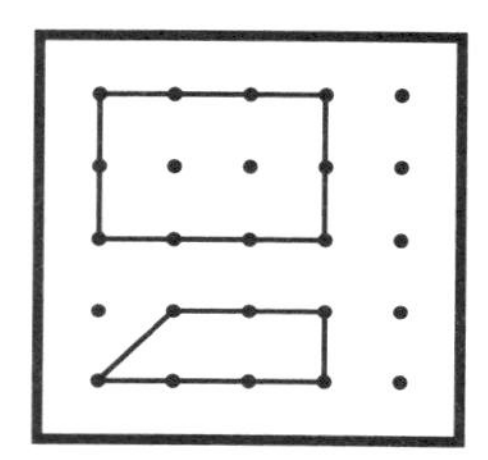

1 square unit of ***area***
Area of the rectangle is 6 square units.
Area of the trapezoid is 2½ square units.

About the Circular Geoboard

The 12-pin circular geoboard is ideal for learning about parts of a circle, central angles of a circle, and angles of inscribed polygons. It also can be used to relate fractions with degrees.

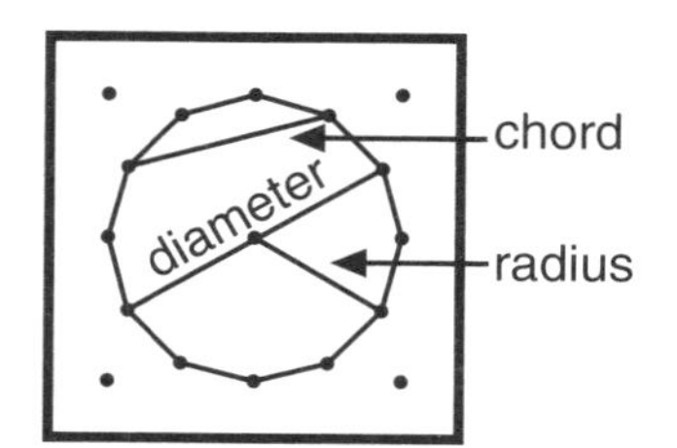

Various parts of a circle can be shown:
diameter, radius, chord

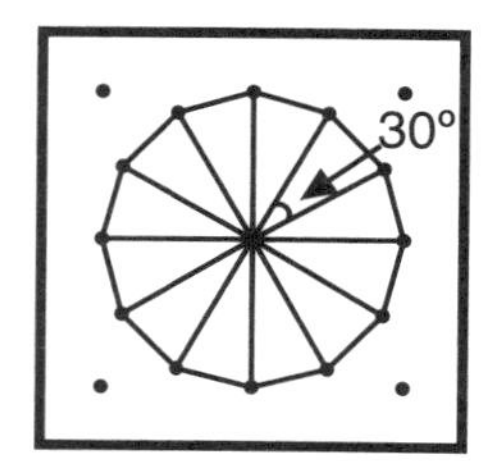

Central angles formed by using all radii show angle measures of 30 degrees each.
A complete rotation is 360 degrees.
Each 30-degree angle is $\frac{1}{12}$ of the circular region.

Lines and Angles

Vocabulary

acute angle, angle, diagonal, line segment, nondiagonal, obtuse angle, parallel lines, perpendicular lines, right angle, vertex

Activity Teaching Notes

Line Segments ***(page 8)***
Students are to model nondiagonal and diagonal line segments on their geoboards.
In Problem 5, before measuring any of the diagonal lines, students use their visual reasoning to rank the diagonal line segments from shortest to longest. Then students compare their measurements to their predictions.

Angles ***(page 9)***
Before students begin this page, have them each make an angle on their geoboards. Display all of them in the chalk tray or on a table. Sort them into groups of ***acute, right,*** and ***obtuse*** angles and then ask students what is common to each group of angles. Samples of acute, right, and obtuse angles are shown at the top of the page, and definitions are provided in the Glossary. Show students how to use an index card to see whether or not an angle is a right angle. Ask each student to model an acute angle and then display some or all of the geoboards on the chalkboard. Discuss the angles. Have students work together to order the acute angles from smallest to largest. Repeat the activity for obtuse angles. If students know how to use a protractor, they may measure the angles to verify their rankings from smallest to largest.

Perpendicular Lines ***(page 10)***
In Problem 1, the lines perpendicular to those shown are relatively obvious. For Problem 2, some students may use an index card to find the lines perpendicular to the ones shown.
As a whole class activity, determine all the different right angles that can be modeled on the geoboards. Children should draw upon their experiences on this sheet to find the right angles in different positions.

Parallel Lines ***(page 11)***
Modeling and drawing the diagonal line segments that are parallel will challenge several children. Since parallel lines usually are shown in pairs in many textbooks and on worksheets, the problems on this page present an opportunity to see several line segments parallel to each other. Ask students to compare their answers to one another.

Name .... Date ....

Line Segments

A ***line*** continues in space forever in both directions.
Line segments are parts of lines with two endpoints. You can model (build) a line segment on the geoboard by connecting any two pins. The pins are the endpoints. You can label a line segment by naming a letter at each of its endpoints.

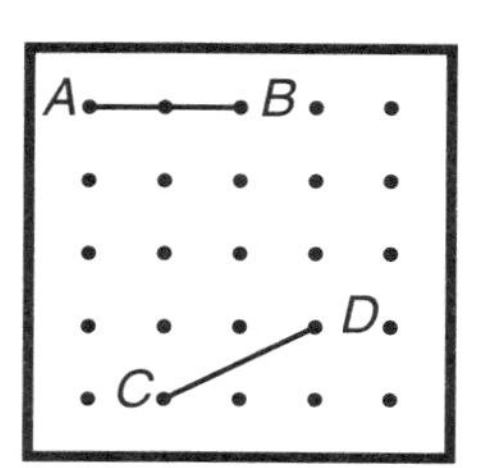

$\overline{AB}$ is nondiagonal.

$\overline{CD}$ is diagonal.

---

1. Model four different nondiagonal lines on your geoboard. Record them below. Label the endpoints *A*, *B*, *C*, etc.

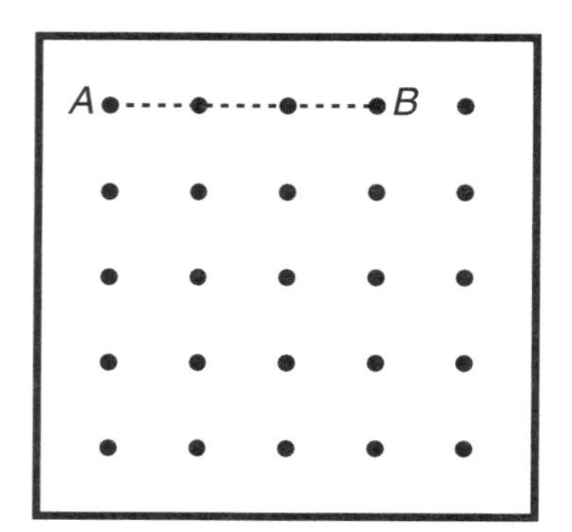

2. If •—• is 1 unit, how long is each nondiagonal line?

$\overline{AB}$ = 3 units ______________________________

____

3. If •—• is 25 millimeters (mm), how long is each nondiagonal line?

$\overline{AB}$ = 75 mm ______________________________

____

4. Model four diagonal lines on your geoboard. Record them below. Label the endpoints.

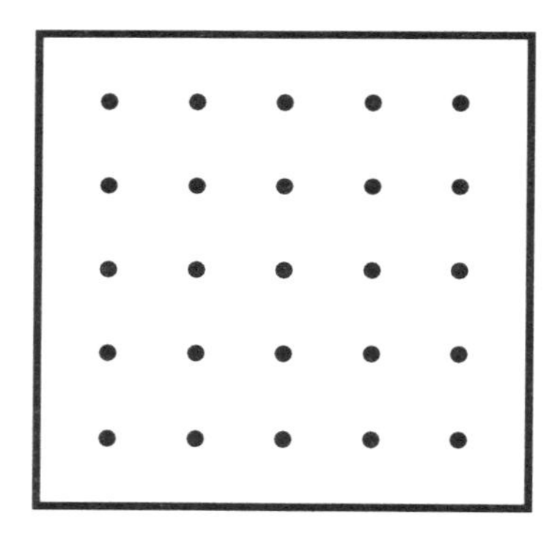

5. Order the diagonal lines in Problem 4 from the shortest to longest without measuring.

____

shortest longest

6. Using a ruler, measure each diagonal line to the nearest millimeter and write the lengths in order below.

____

Name .... Date ....

Angles

When two line segments on your geoboard meet at a common endpoint, they form an ***angle.*** The common endpoint is called the ***vertex.***

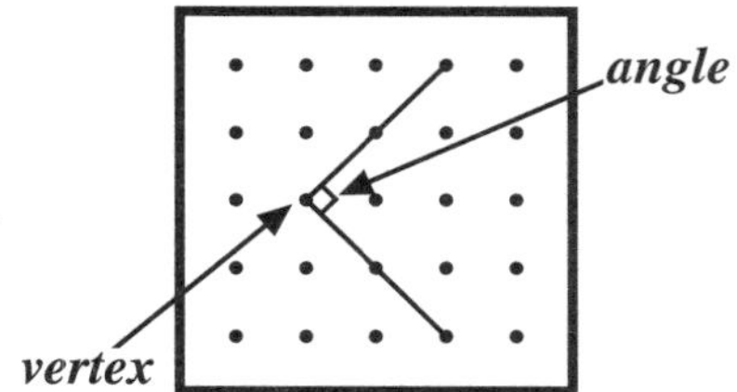

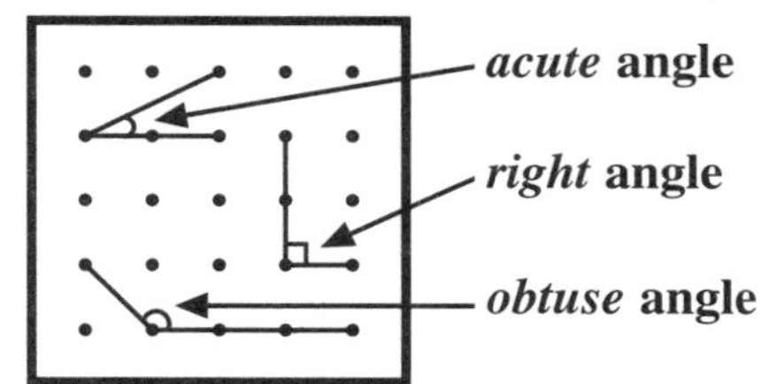

1. Model three different acute angles on your geoboard.
Record the acute angles on the geoboards below.
Starting with 1, number each of your acute angles from smallest to largest.

a.

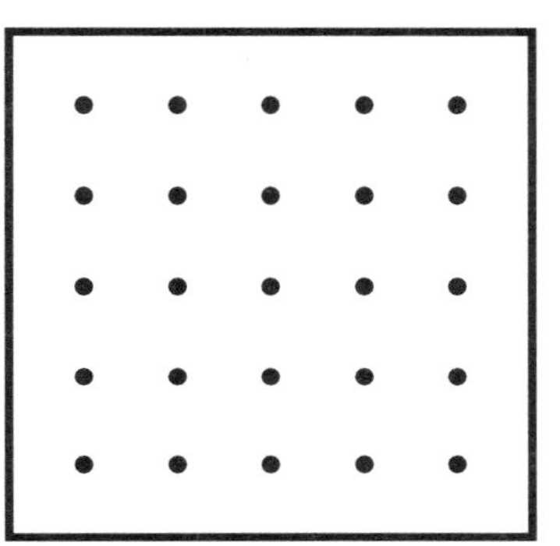

b.

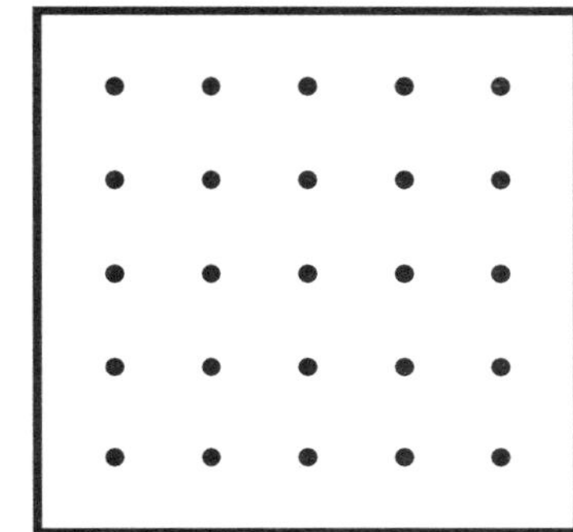

c.

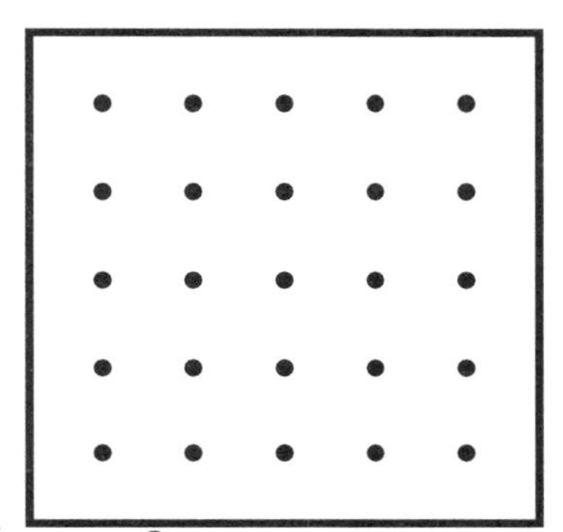

2. Model three different obtuse angles on your geoboard.
Record them below. Starting with 1, number each of your obtuse angles from smallest to largest.

a.

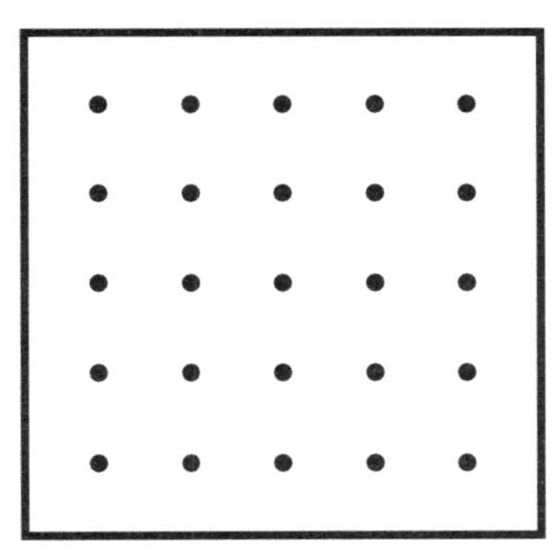

b.

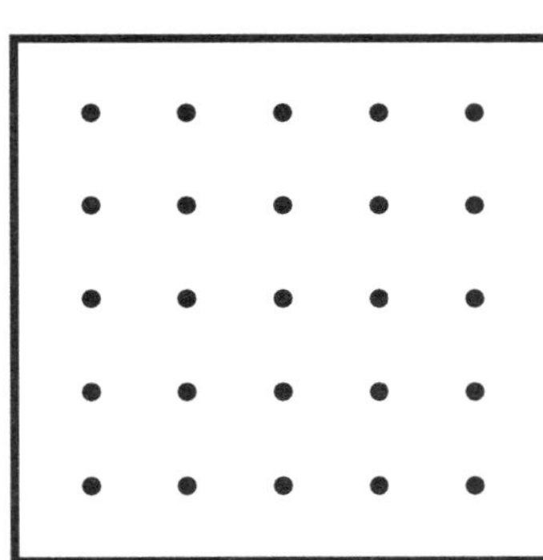

c.

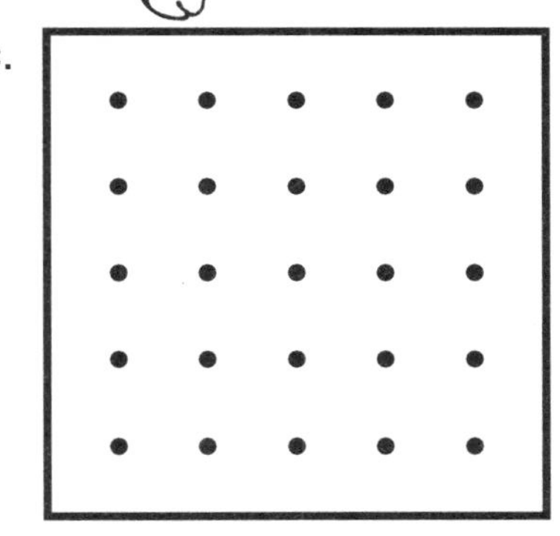

3. How could you be sure the angles you recorded above are not right angles?

Name .... Date ....

Perpendicular Lines

Perpendicular lines intersect at right angles.

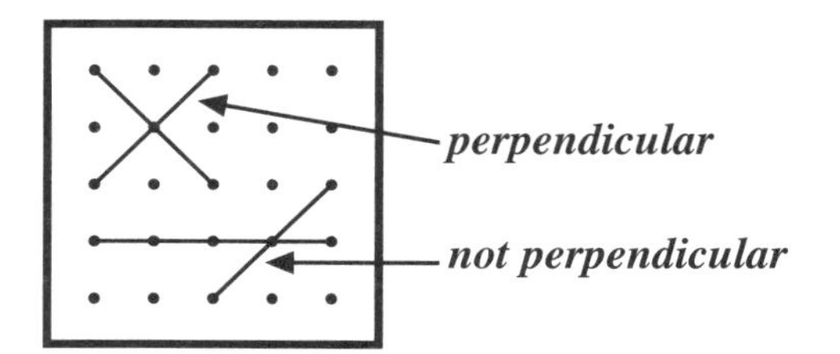

1. Model three different line segments perpendicular to each line segment shown. Record them.

a. 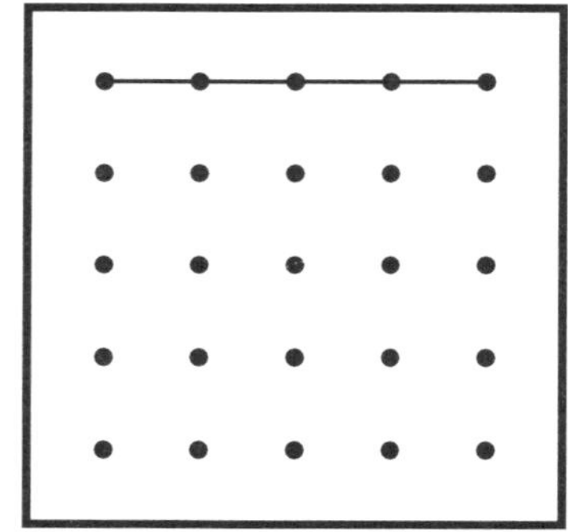

b. 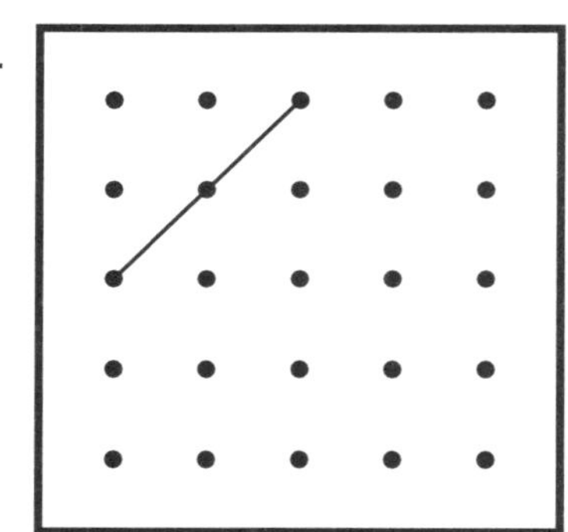

c. 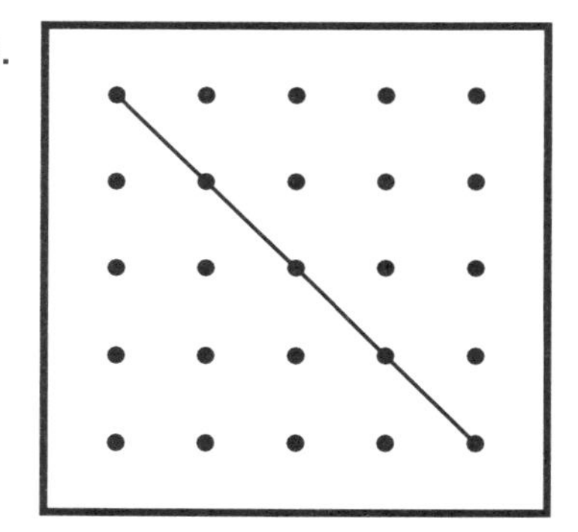

2. Model three different line segments perpendicular to each line segment shown. Record them.

a. 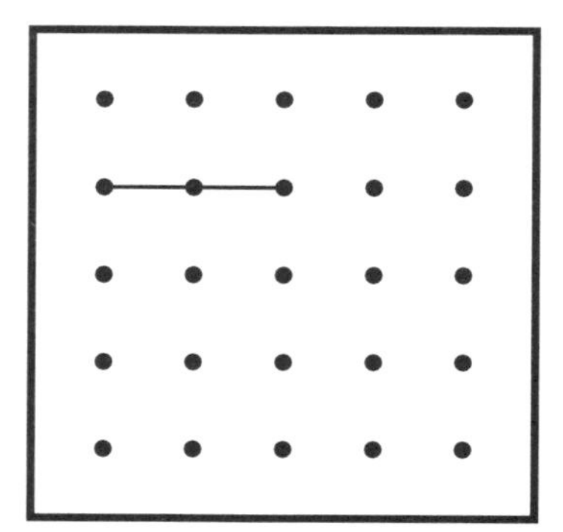

b. 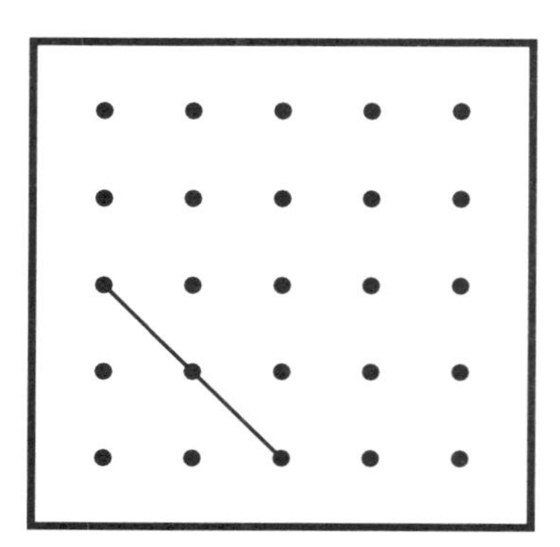

c. 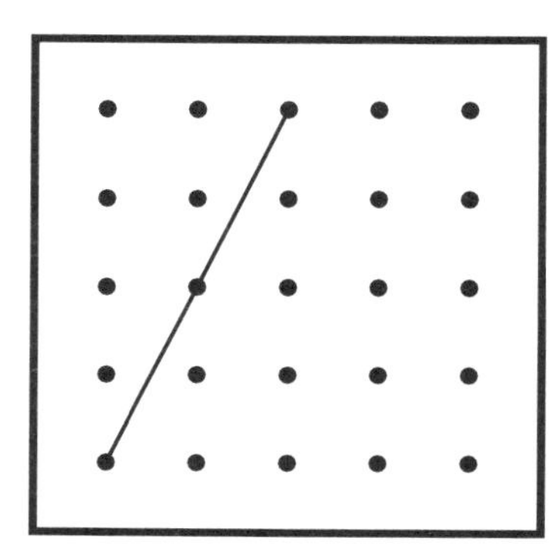

Name .... Date ....

Parallel Lines

Parallel lines are lines in the same plane that never meet. They are always the same distance apart, and continue forever in both directions.

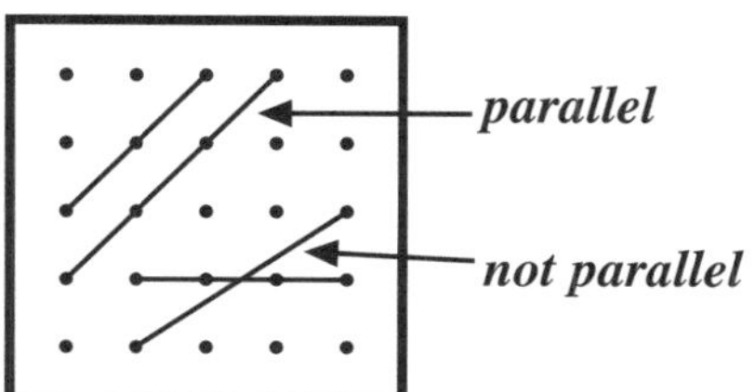

1. Model four different line segments on your geoboard parallel to each of the line segments shown, and record them.

a.

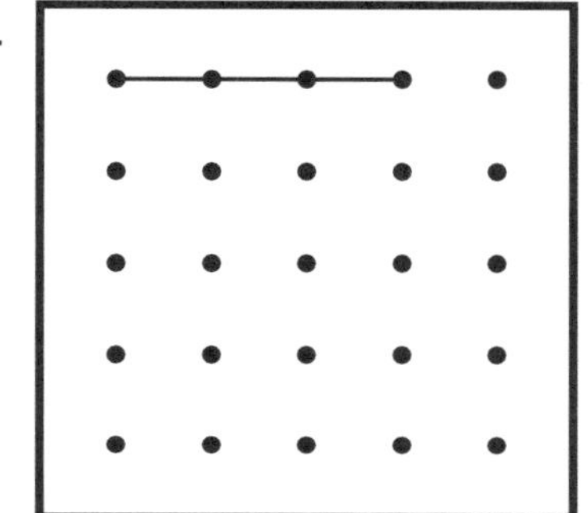

b.

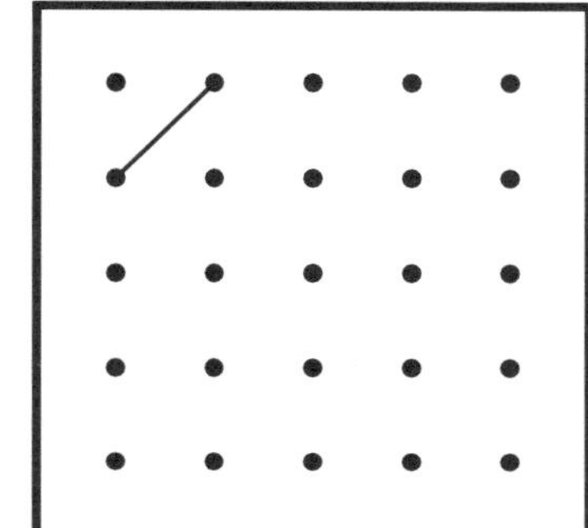

c.

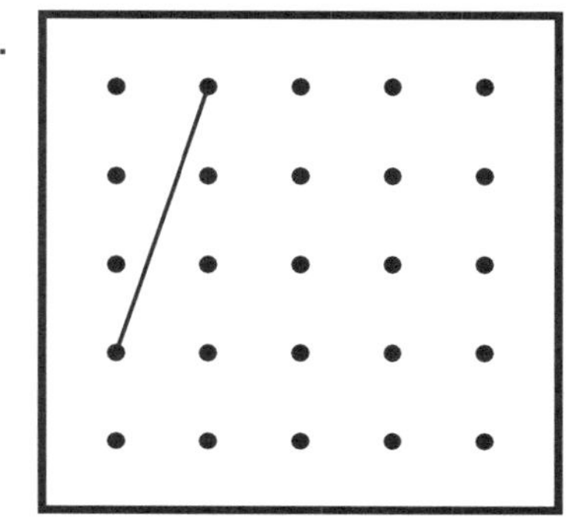

2. Model the longest pair of parallel line segments on your geoboard. Record them.

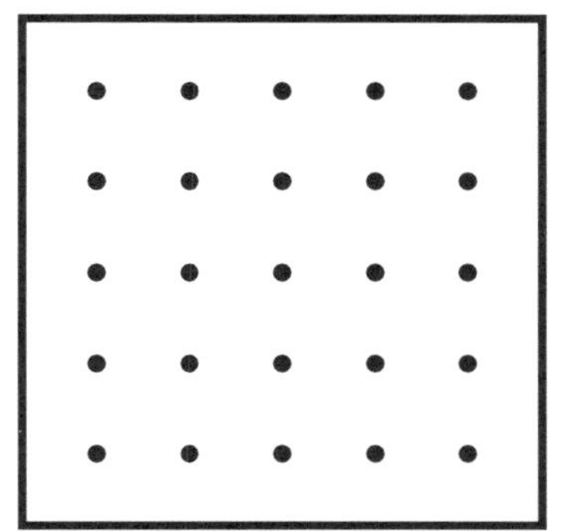

3. Model the shortest pair of parallel line segments on your geoboard. Record them.

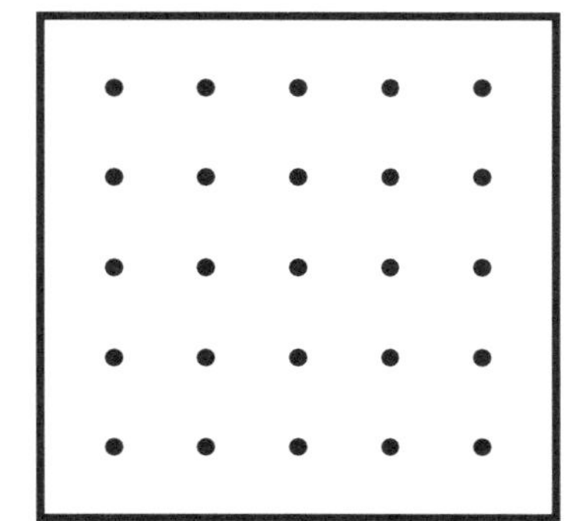

4. Show three or more line segments parallel to each other on your geoboard. Record them.

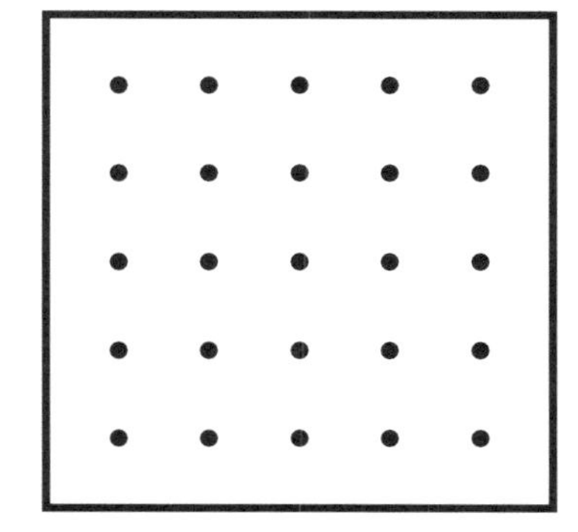

Polygons & Geometric Relationships

Vocabulary

acute triangle, congruent, equilateral triangle, hexagon, isosceles triangle, kite, obtuse triangle, octagon, parallelogram, pentagon, polygon, quadrilateral, reflection (flip), regular polygon, rhombus, rotation (turn), similar, symmetric, translation (slide), trapezoid

Activity Teaching Notes

Polygons ***(page 14)***
Have students show their polygons to other classmates and discuss them. Students are to label their recorded models as "regular" polygons if they have sides of equal lengths and congruent angles. For Problem 2, have students make a list of names for all the different polygons according to their number of sides. Refer to the Glossary.

Triangles ***(page 15)***
Six types of triangles are defined in terms of sides ***(equilateral, isosceles,*** and ***scalene)*** and angles ***(acute, right,*** and ***obtuse).*** Although students are asked only to model and record two examples of each type of triangle, give each of them several recording sheets to find as many of each triangle as they can. Ask students to work in cooperative groups.

More Triangles ***(page 16)***
In this activity, triangles are classified by two attributes and named using two words, such as ***"isosceles right triangle."*** Challenge students to find more models of each triangle.

Quadrilaterals ***(page 17)***
Children model six different quadrilaterals in this activity. Have students talk about their descriptions for each quadrilateral. Then ask them to write their own definitions for each figure on the backs of their sheets.

Similar Squares ***(page 18)***
This activity combines the concept of similarity with the task of modeling eight different squares on the geoboard. Make sure students check to see that there are four right angles and four sides of equal length for each of the squares.

Congruent Figures ***(page 19)***
This activity focuses on the concept of congruence and reviews different types of quadrilaterals. Provide extra recording sheets for students to model more figures congruent to the ones shown.

Other Polygons ***(page 20)***
Students model pentagons, hexagons, and octagons using a specified number of pins for Problems 1-6. Then they are asked to model more figures and to determine whether or not they are congruent or similar to other figures. Extend this activity by providing students with several copies of the recording sheets. Challenge them to find as many pentagons, hexagons, and octagons as possible on a 5 x 5-geoboard.

Symmetric Figures ***(page 21)***
Tell students to use one rubber band to model each polygon and another rubber band to show the line of symmetry. If there are several lines of symmetry, have them show another with additional rubber bands. Ask students if there are pentagons, hexagons, trapezoids, or other quadrilaterals that do not have any lines of symmetry. Discuss which types of triangles, quadrilaterals, and other polygons have 0, 1, 2, or more lines of symmetry. You may want to provide small mirrors so students can see the mirror image of one side of a symmetric figure.

Slides, Flips, and Turns ***(page 22)***
Single moves are given in Problems 1-3; double moves are given in Problems 4-6. Ask students which types of moves are made. Ask whether the figures are still congruent after each move. One of the advantages of using the geoboard to show a rotation is that the student can model a figure and then physically turn the geoboard to see how the moved figure will look. Also, students can use small mirrors to help them find the reflection of a figure about a line. Urge students to create motion problems to challenge each other.

Name .... Date ....

Polygons

A ***polygon*** is a simple closed figure made up of straight line segments. The line segments are called ***sides.***
A ***vertex*** is the common endpoint where two sides meet.

In a ***regular polygon,*** all the sides are the same length and all angles are congruent.

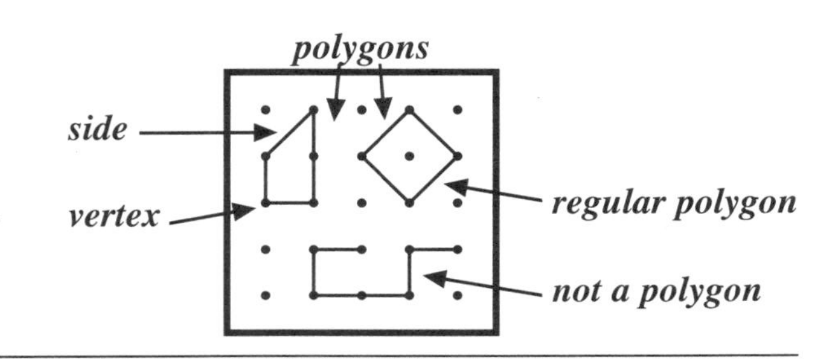

1. Model five different polygons. Record them below.
Write "regular polygon" near any that are regular polygons.

a.

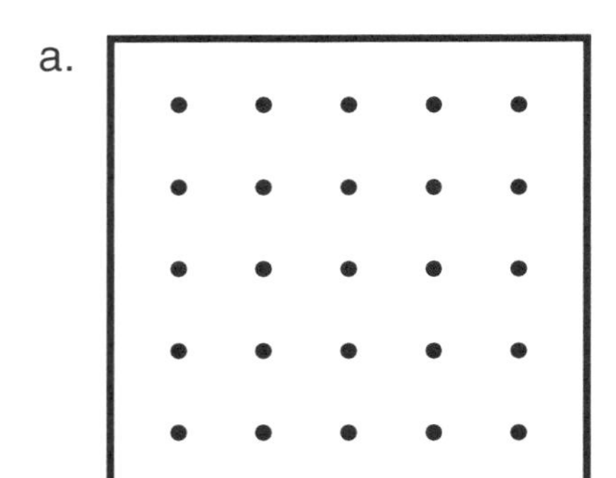

b.

c.

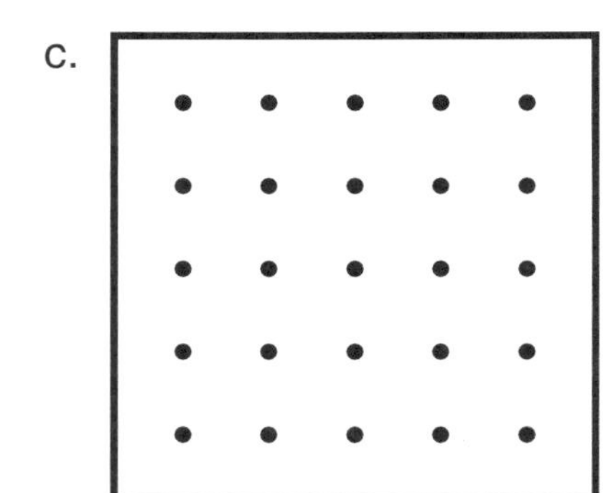

d.

e.

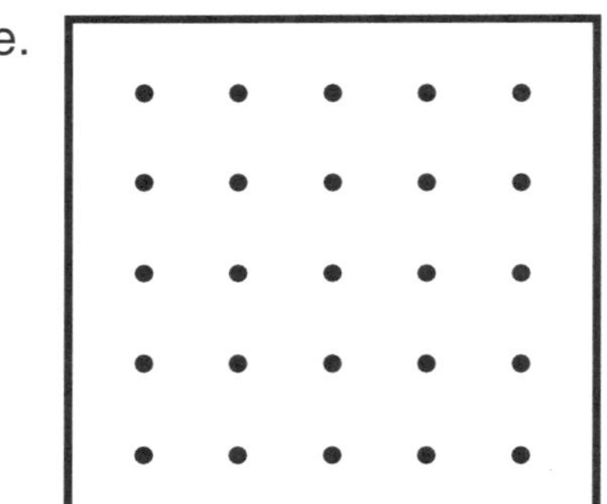

2. Can you name any of the polygons you made?
Make a list of polygon names.
Name each polygon above.

Name .... Date ....

Triangles

A ***triangle*** is a polygon with 3 sides.

Triangles can be named by their sides.

Equilateral: all sides same length

Isosceles: two sides same length

Scalene: all sides different lengths

Triangles can be named by their angles.

Acute: has all acute angles

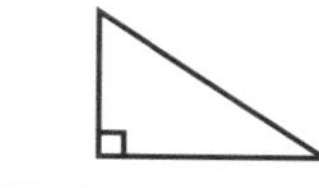

Right: has a right angle

Obtuse: has an obtuse angle

Model two examples of each type of triangle on your geoboard. Record them below. One type of triangle cannot be made with the geoboard. Which type?

1. Scalene triangle

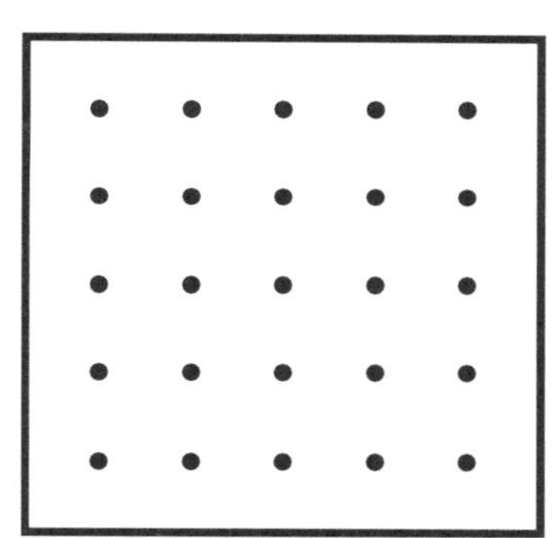

2. Equilateral triangle

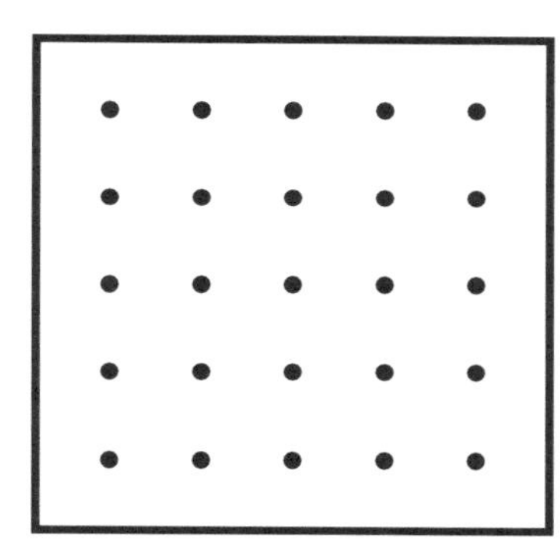

3. Isosceles triangle

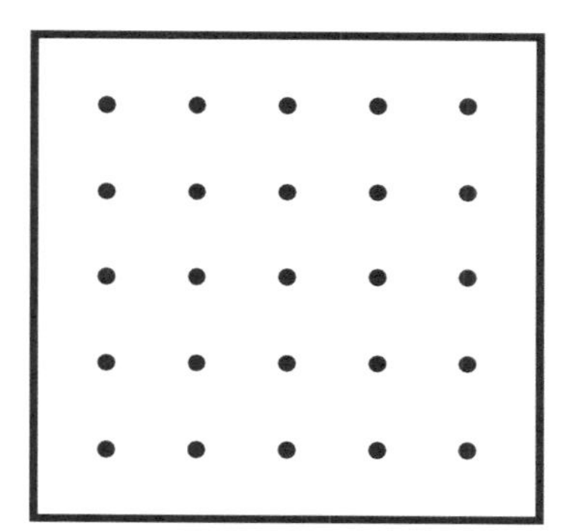

4. Right triangle

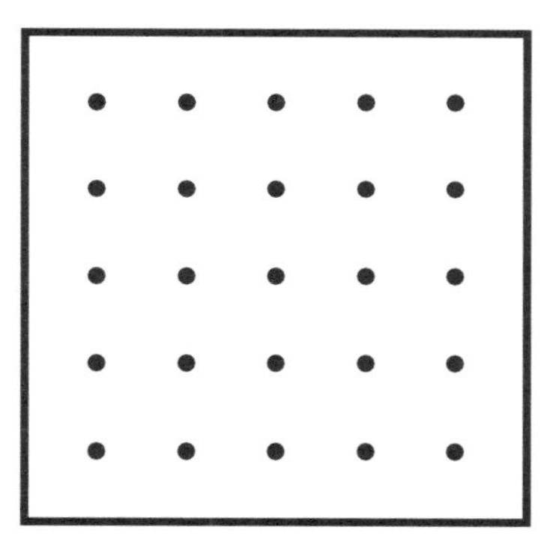

5. Acute triangle

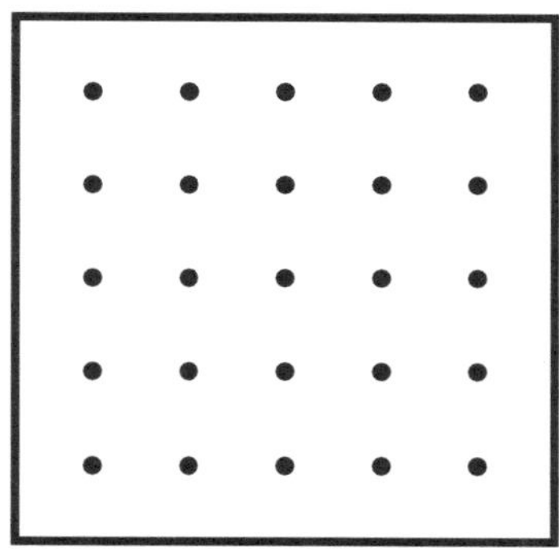

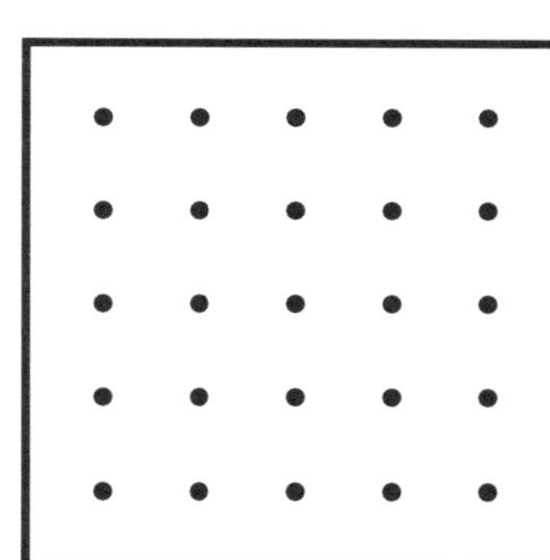

6. Obtuse triangle

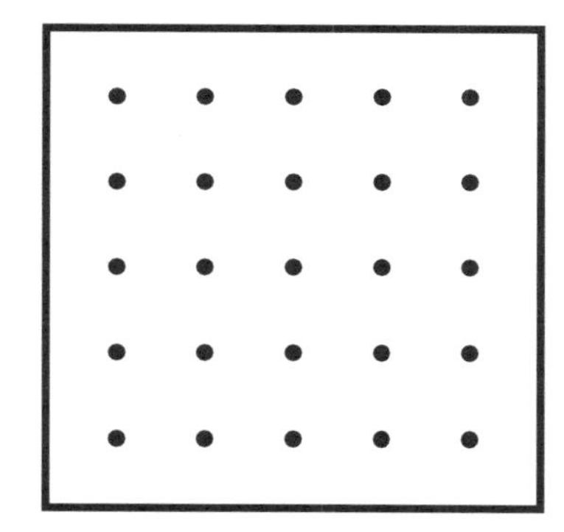

Name .... Date ....

More Triangles

Triangles may be named using two words.

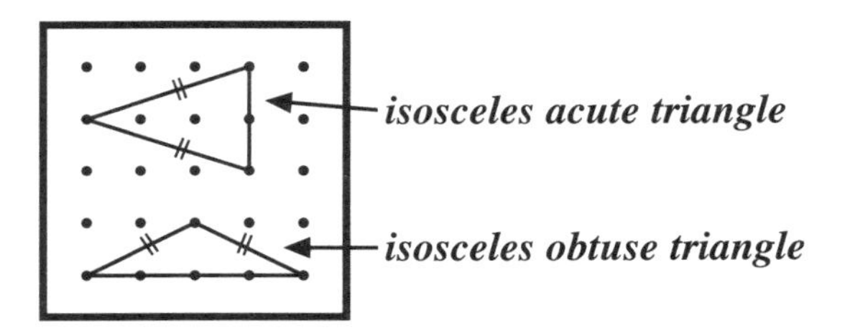

Model two different examples of each type of triangle on your geoboard. Record them below.

1. Isosceles right triangle

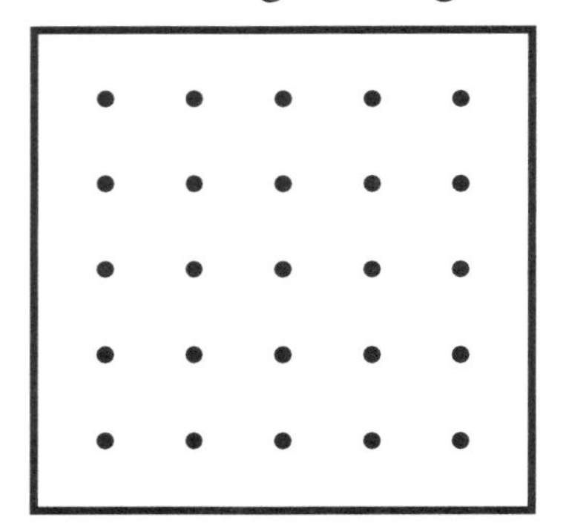

2. Scalene acute triangle

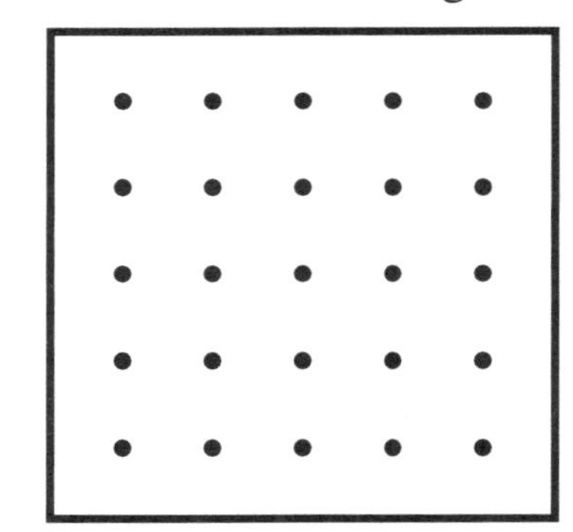

3. Scalene right triangle

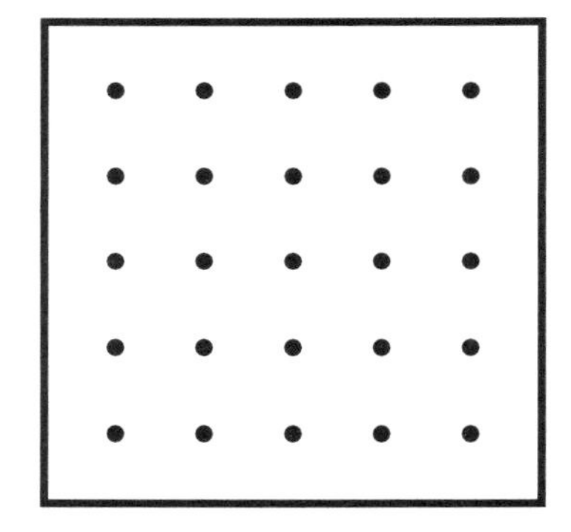

4. Isosceles acute triangle

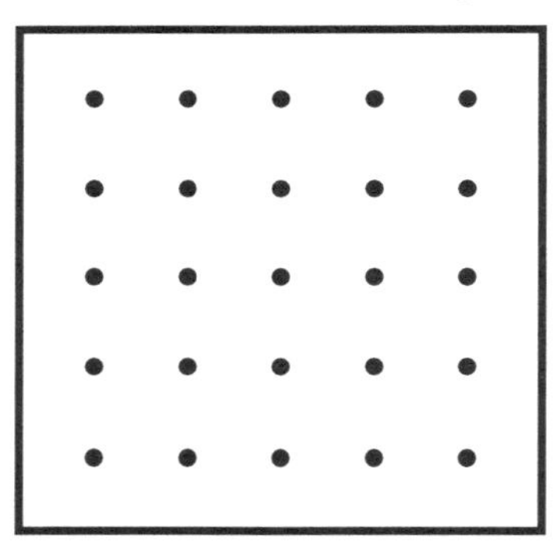

5. Scalene obtuse triangle

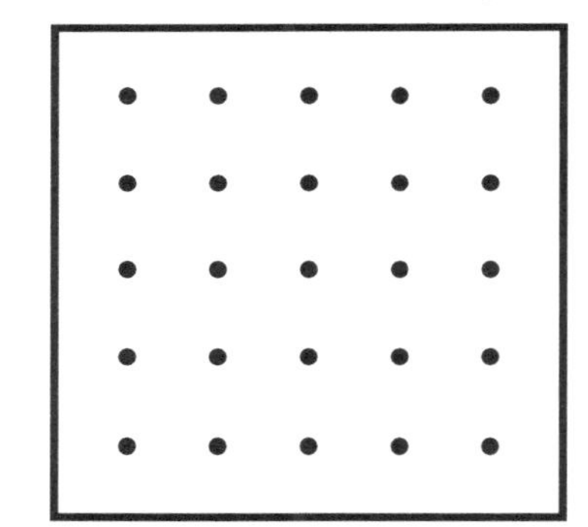

6. Isosceles obtuse triangle

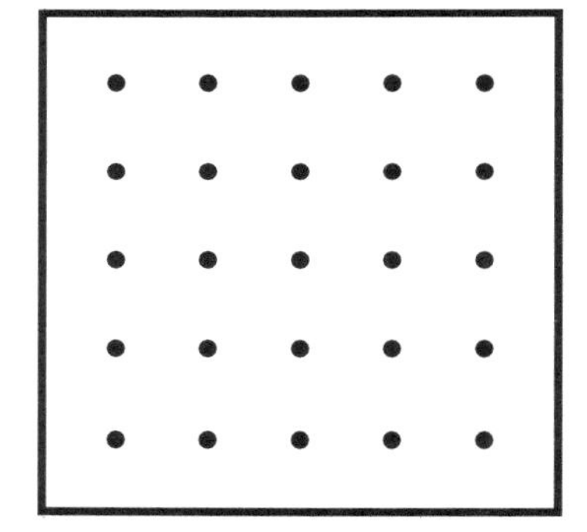

Name the following triangles using two words.

7. ____________________

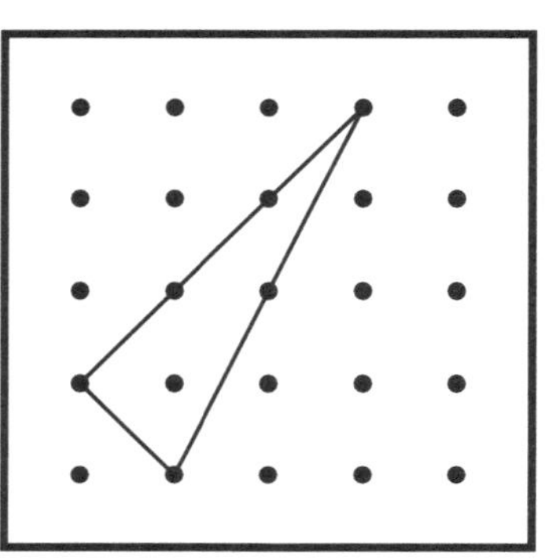

8. ____________________

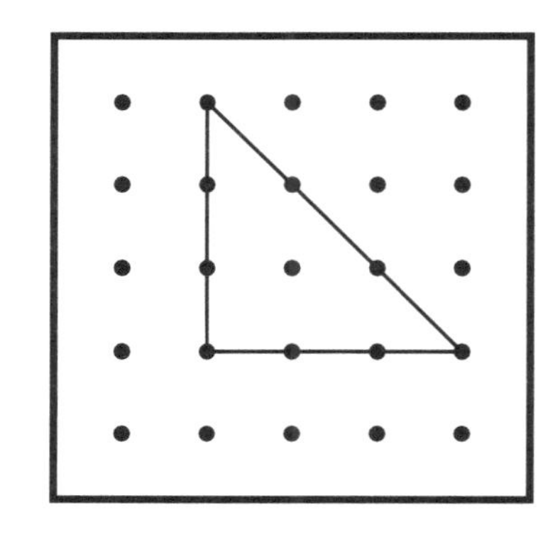

Name .... Date ....

Quadrilaterals

A ***quadrilateral*** is a polygon with 4 sides.

Some have ***right angles.***
Some have ***sides with equal lengths.***
Some have ***pairs of opposite sides parallel.***

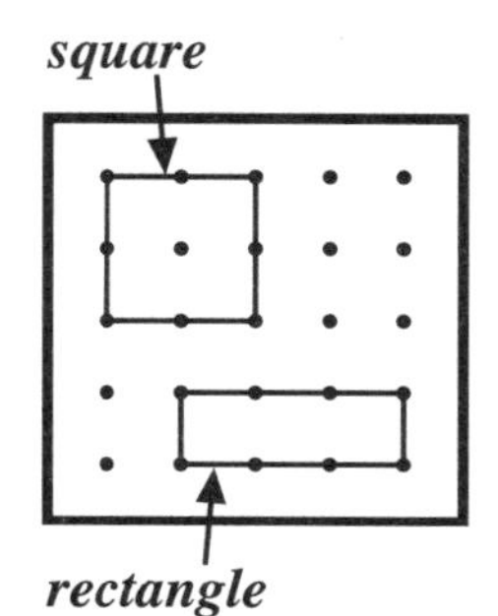

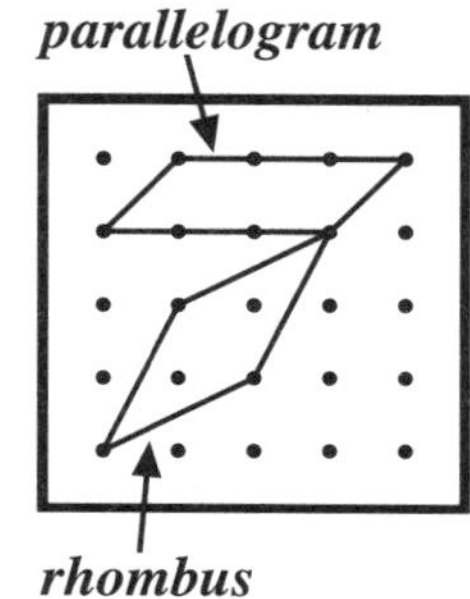

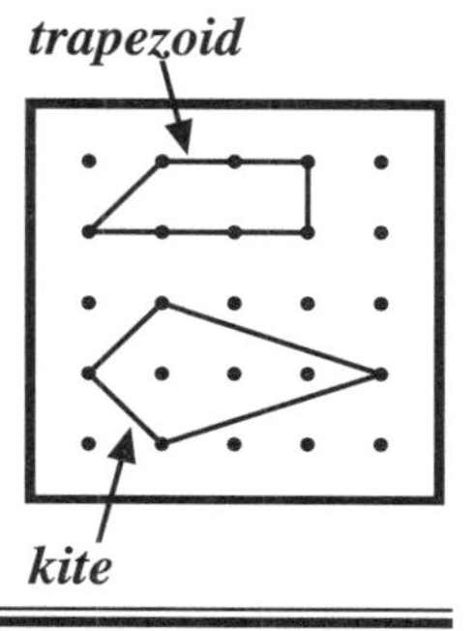

Model two examples of each quadrilateral on your geoboard. Record them below.

1. Rectangle

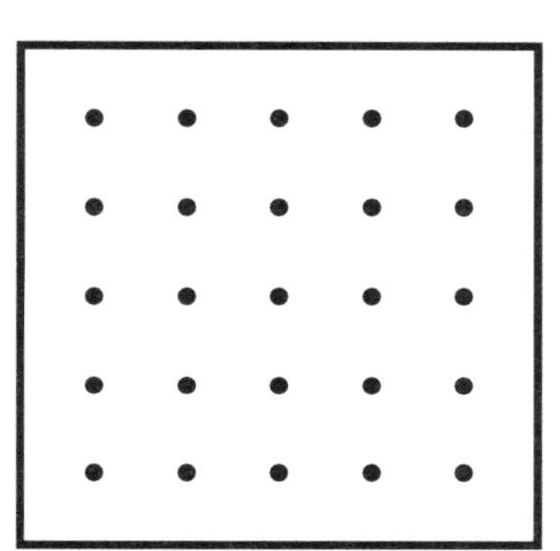

2. Square

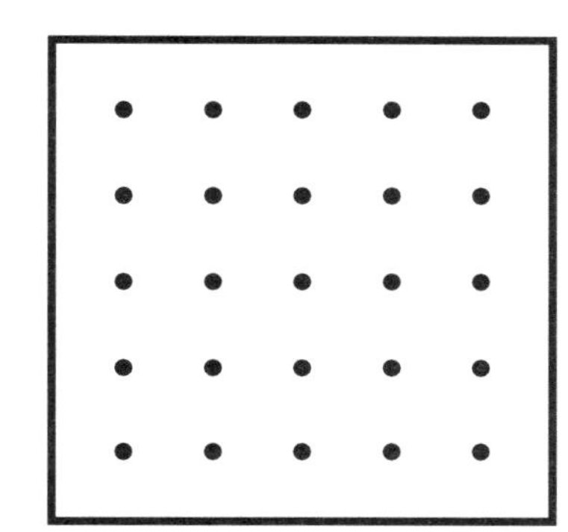

3. Parallelogram

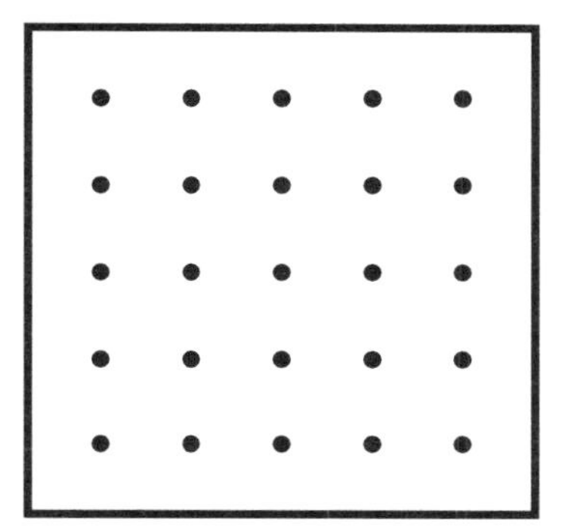

4. Trapezoid

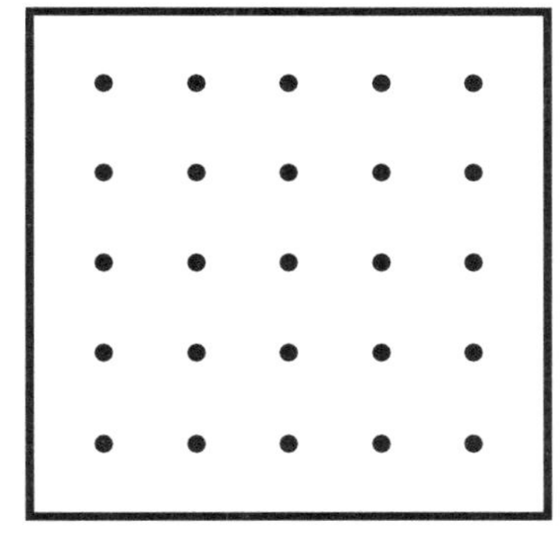

5. Kite

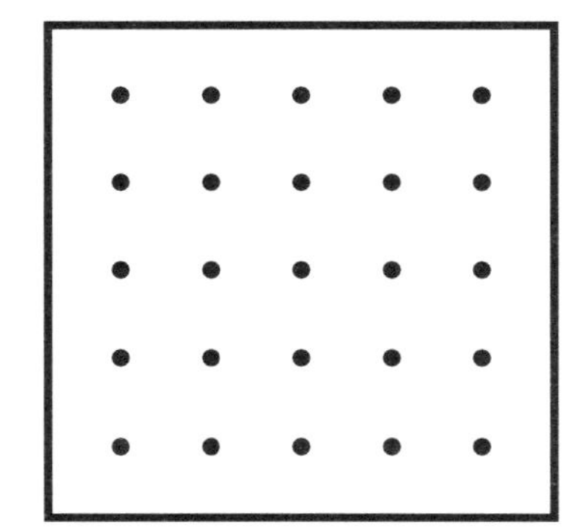

6. Rhombus

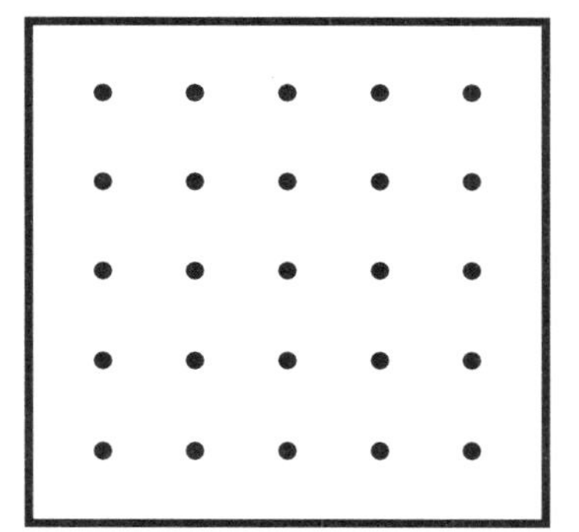

7. On the back of this sheet, describe each of the different types of quadrilaterals. Include observations about their angles and side lengths.

Name .... Date ....

Similar Squares

Similar figures have the same shape and corresponding proportional sides. A similar shape is like an enlargement or a reduction of a shape.

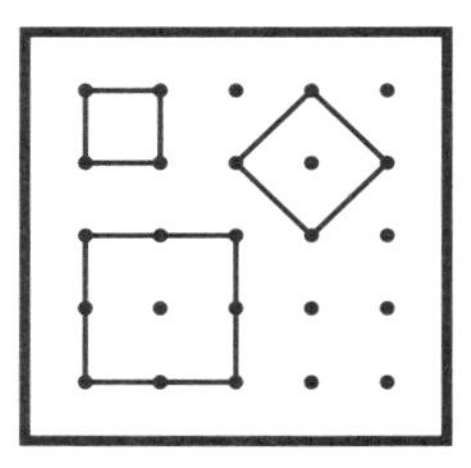

Model eight similar squares on a 5 x 5 geoboard. Record each one.

1.

2.

3.

4.

5.

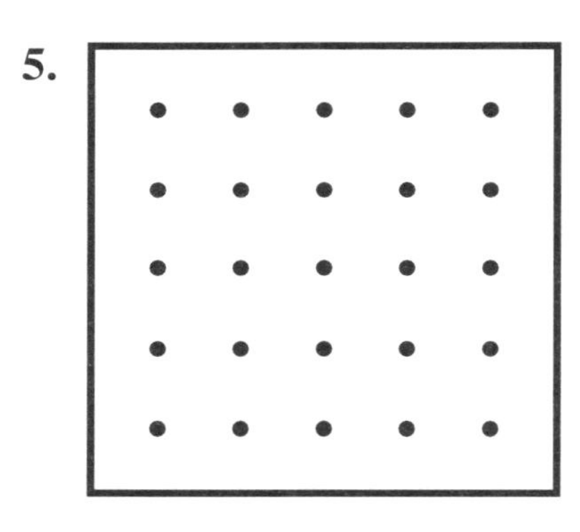

6.

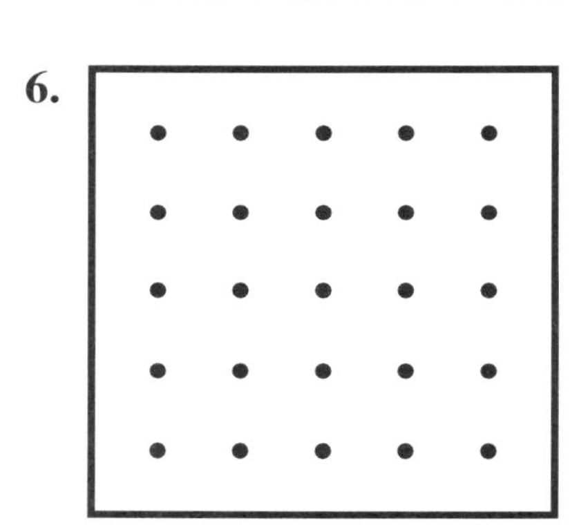

7.

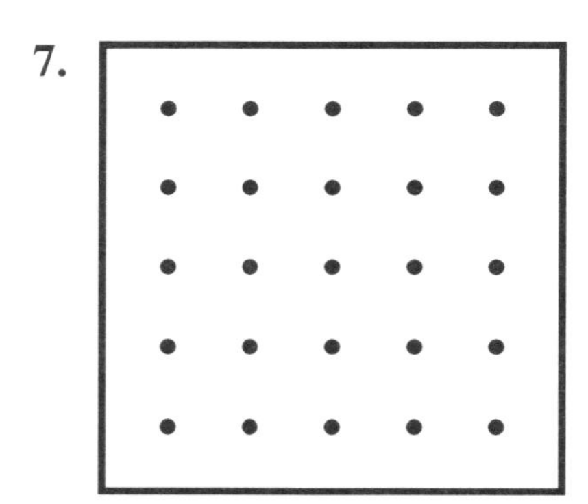

8.

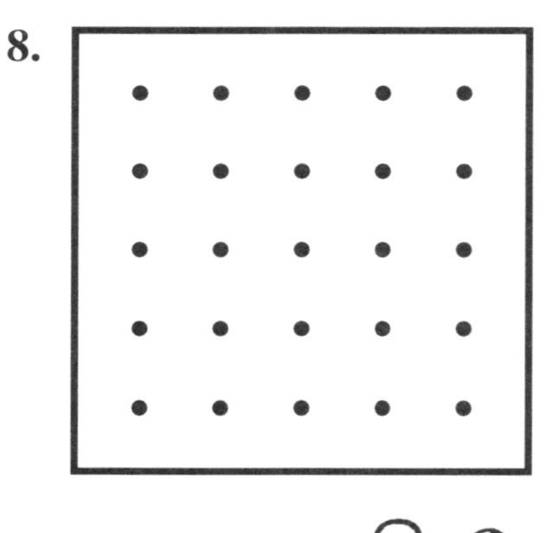

Check each square:
Are all the angles right angles?
Are all the sides the same size?
Are pairs of opposite sides parallel?

Name .... Date ....

Congruent Figures

Congruent figures have the same size and shape.

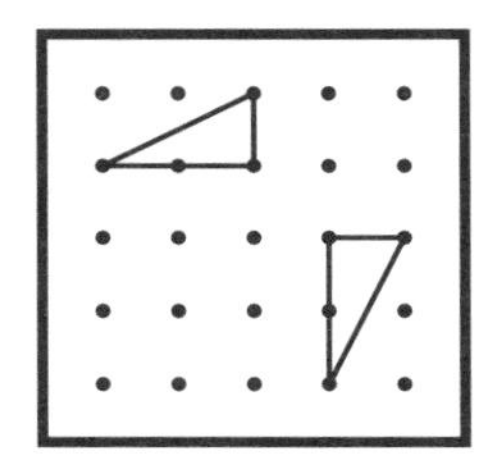

Use two rubber bands. Model the figure shown with one rubber band. Then model a new figure congruent to the first and record it. Name the shape of each figure.

1. ____________________ 2. ____________________ 3. ____________________

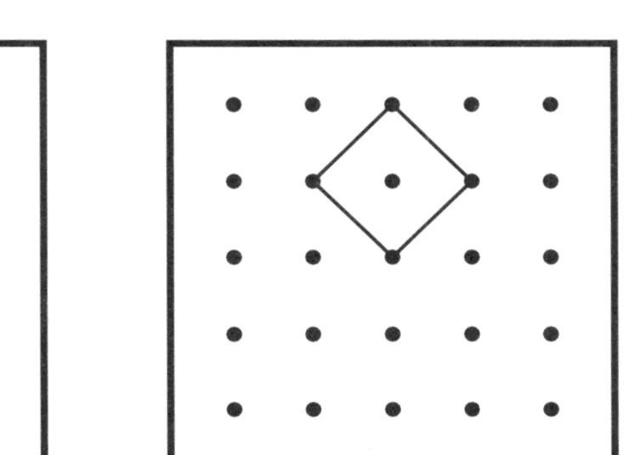

4. ____________________ 5. ____________________ 6. ____________________

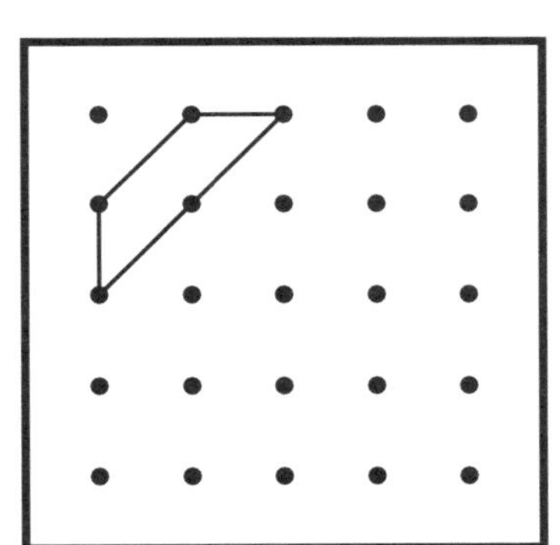

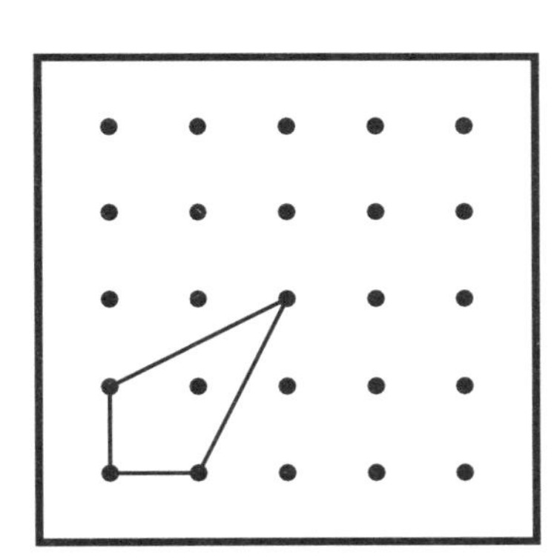

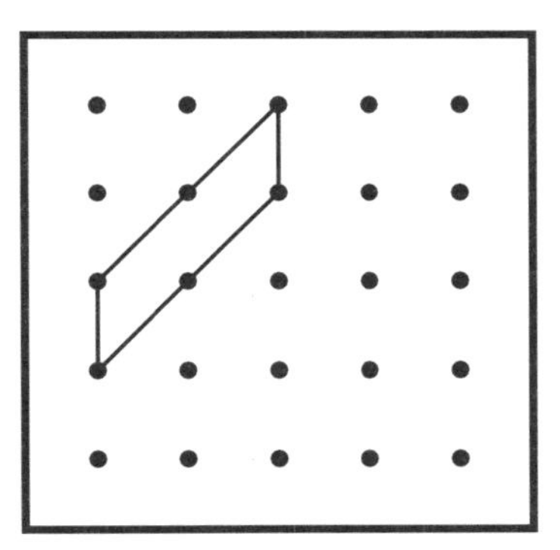

7. Which other types of figures could you model with a pair of congruent figures? Try it on a geoboard.

Name .... Date ....

Other Polygons

Study the chart of polygon names and the number of sides.

Polygon	Number of sides
Pentagon	5
Hexagon	6
Octagon	8

Model each polygon touching a specified number of pins. Record them below.

1. Hexagon; 6 pins

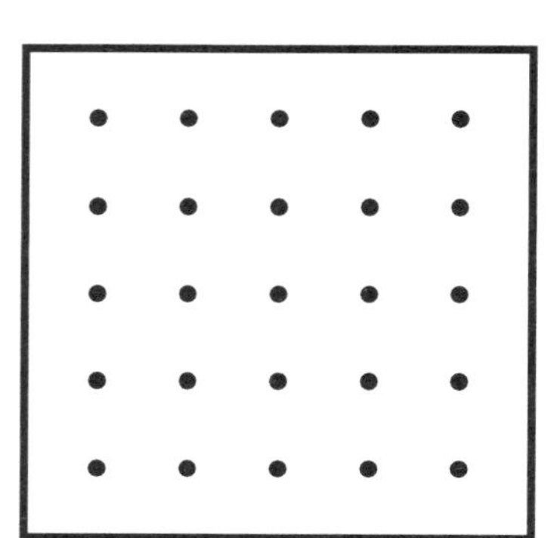

2. Pentagon; 5 pins

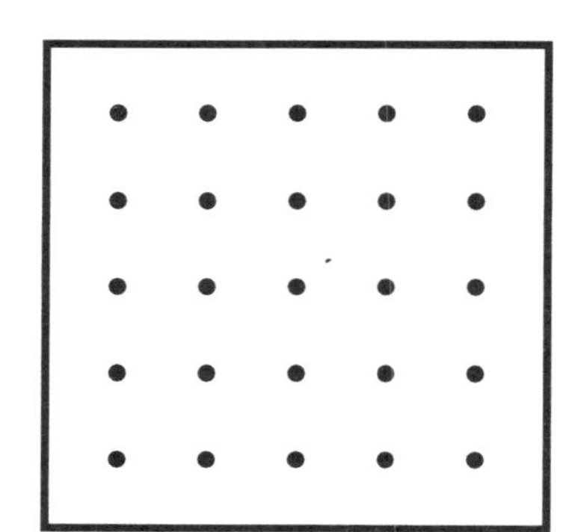

3. Octagon; 8 pins

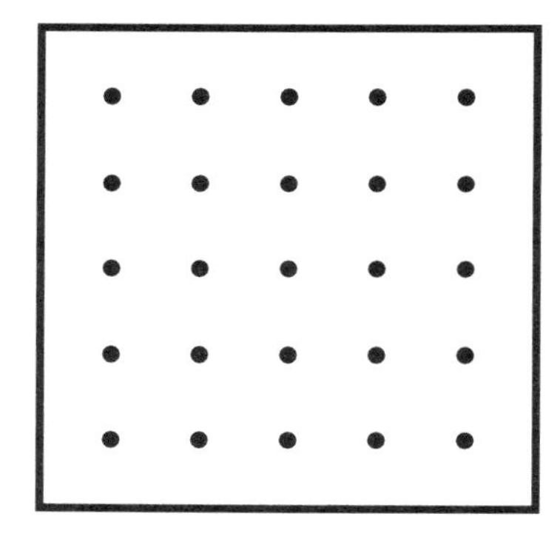

4. Hexagon; 14 pins

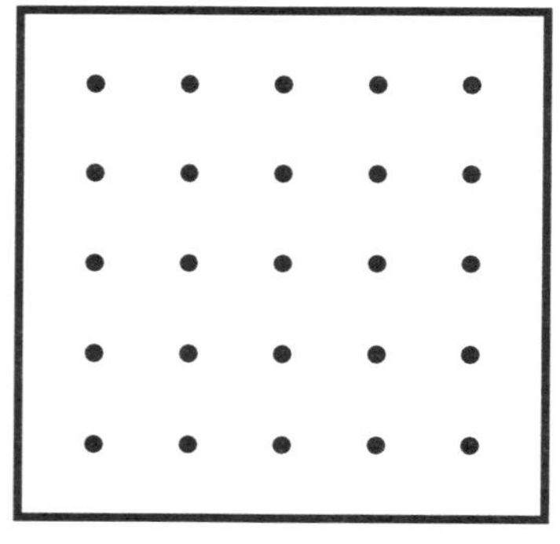

5. Octagon; 12 pins

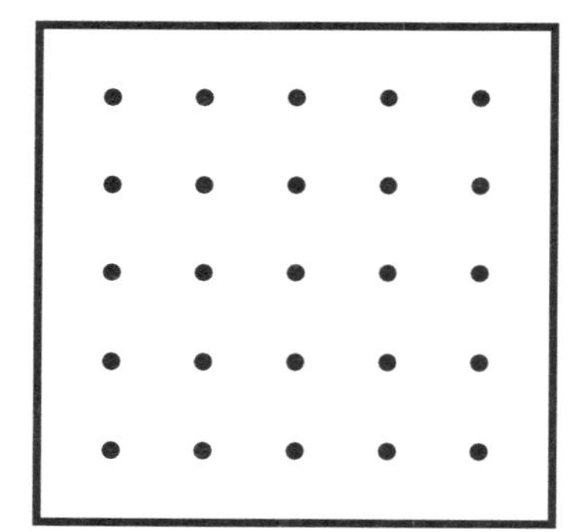

6. Pentagon; 9 pins

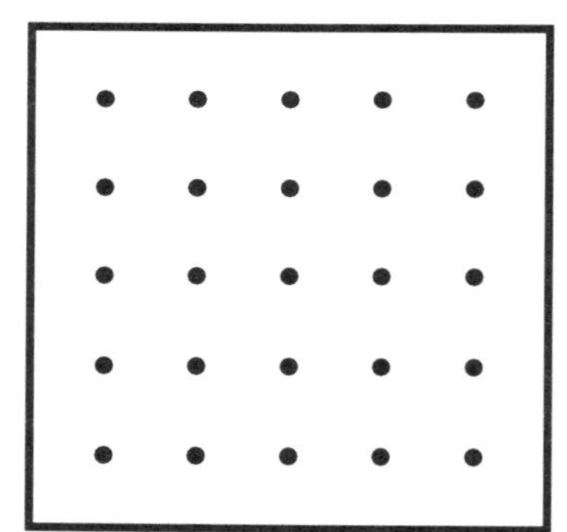

7. For Problem 1, model and record another hexagon touching 6 pins. Are they congruent? Are they similar?

8. For Problem 2, model and record a pentagon touching 10 pins. Are they congruent? Are they similar?

Name .... Date ....

Symmetric Figures

A ***symmetric*** figure has at least one ***line of symmetry*** which separates it into two identical halves. One side is a ***reflection,*** or mirror image, of the other side.

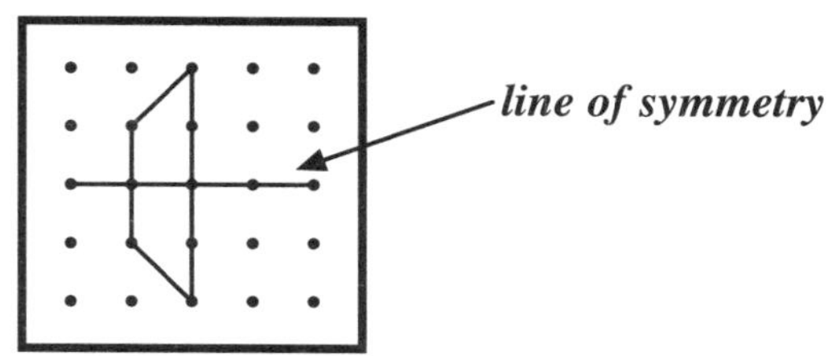

Use two or more rubber bands for each problem. Model a polygon with one rubber band. Use the other rubber band(s) to show the line(s) of symmetry. Record them below.

1. Pentagon;
1 line of symmetry

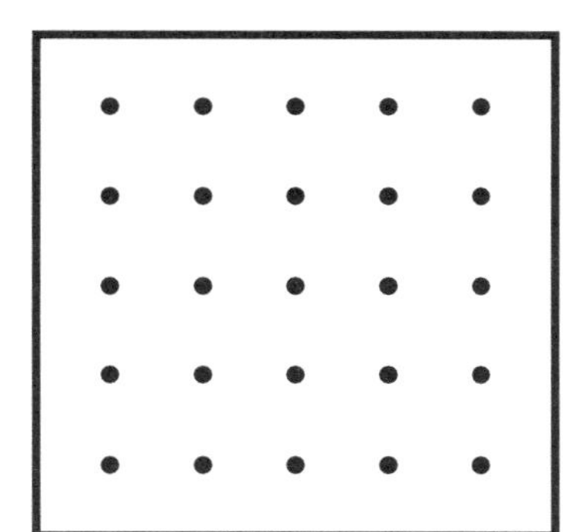

2. Trapezoid;
1 line of symmetry

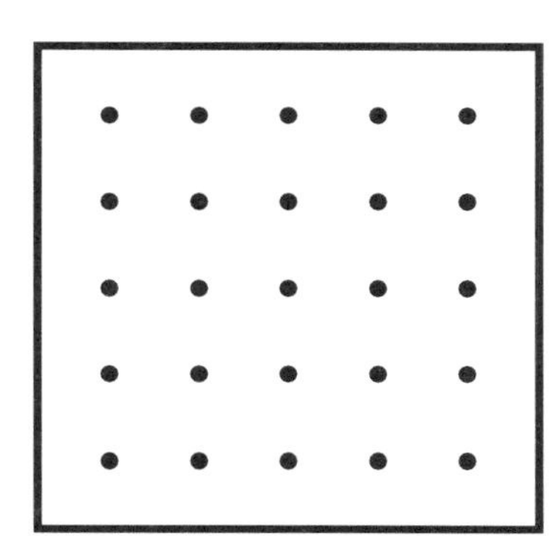

3. Triangle;
0 lines of symmetry

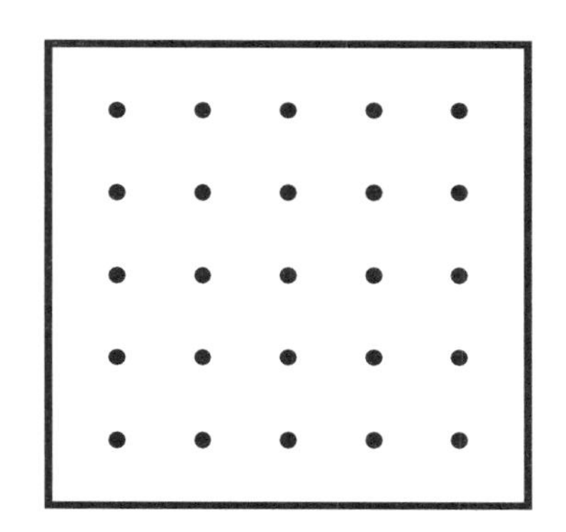

4. Triangle;
1 line of symmetry

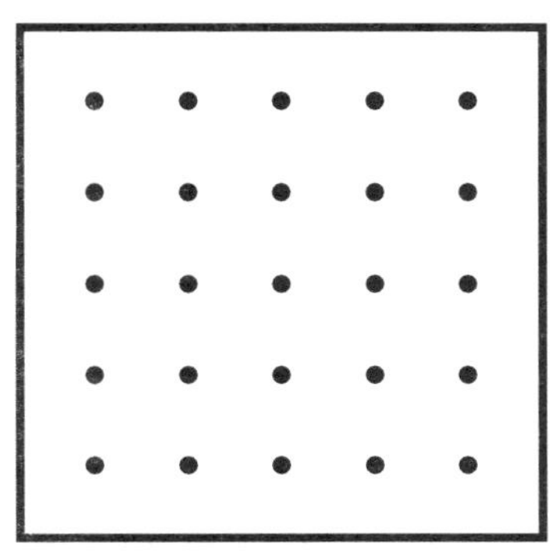

5. Hexagon;
2 lines of symmetry

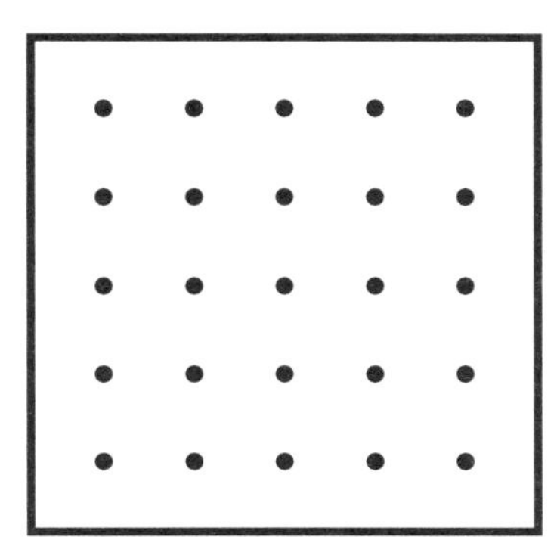

6. Quadrilateral;
4 lines of symmetry

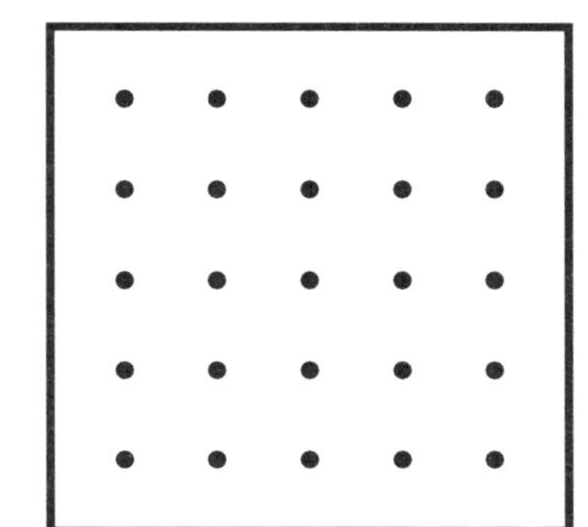

7. Which type of triangle has 1 line of symmetry? ________________

8. Which type of quadrilateral has exactly 2 lines of symmetry? ________________

9. Which types of quadrilaterals could have 0 lines of symmetry? _______, _______, _______

10. What is the name of the quadrilateral for Problem 6? ________________

Name .... Date ....

Slides, Flips, and Turns

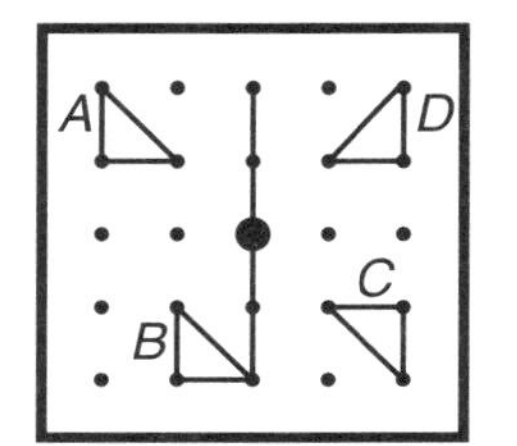

Figures can be moved in different ways and not lose their size or shape. Look at Figure A.

A ***translation,*** or slide, is a motion involving distance and direction.
Figure B moved 1 across and 3 down from Figure A.

A ***rotation,*** or turn, is a motion of a figure around a fixed point in a certain direction.
Figure C was moved ½ turn to the right of Figure A, clockwise around the center point of the geoboard.

A ***reflection,*** or flip, is a motion about a line to show a mirror image.
Figure D is a mirror image of Figure A about the line.

Model each figure on your geoboard and record it after the motion. Write the type of motion it is in the blank.

1. Move 3 left and 2 up.

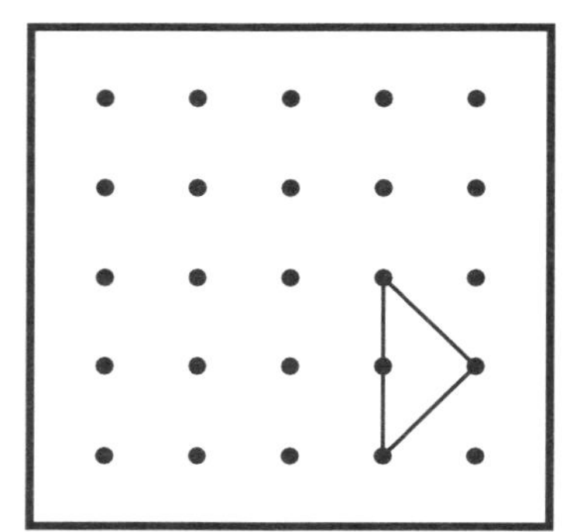

____

2. Move about the line.

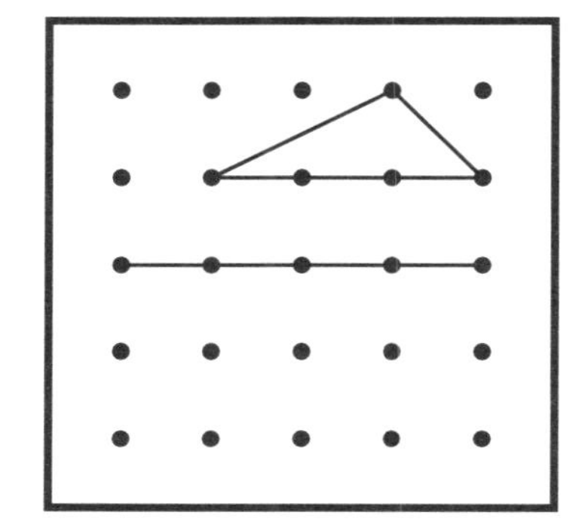

____

3. Move ¼ turn counterclockwise.

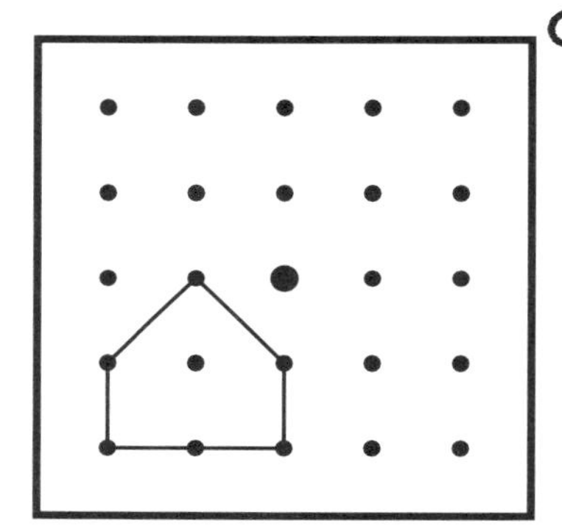

____

4. Move about the line, then move 1 right and 2 down.

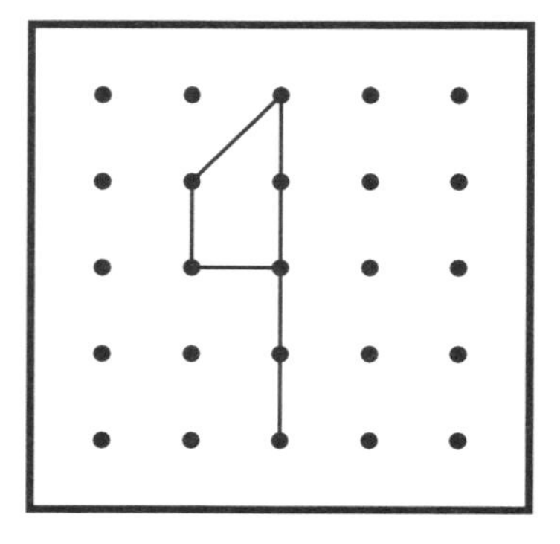

____

5. Move ½ turn clockwise, then 1 left and 2 up.

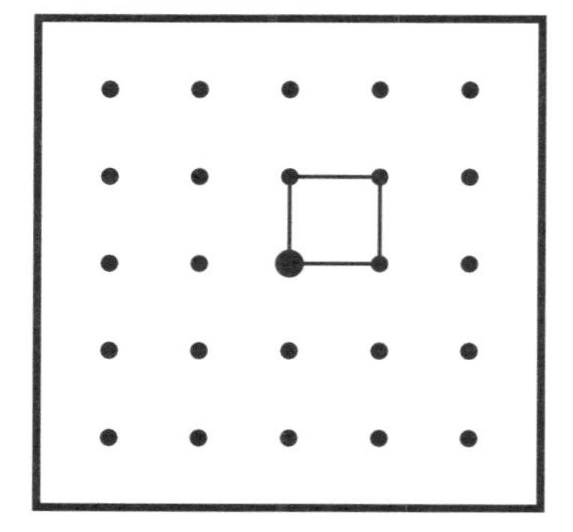

____

6. Move 3 right and 1 down, then 2 left and 2 up.

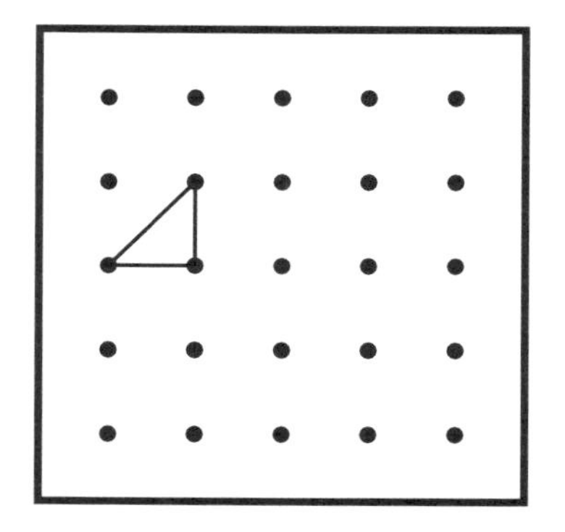

____

Perimeter and Area

Vocabulary

area, perimeter, square unit, unit

Activity Teaching Notes

Perimeter ***(page 24)***

This activity deals with the perimeter of squares and rectangles where only whole units of length are used. Make sure students know they can add the lengths of all the sides to find the perimeter. Some also may know that the perimeter of a square is 4 times the length of one side ***(P = 4s)*** and the perimeter of a rectangle is the sum of 2 lengths and 2 widths ***(P = 2l + 2w).*** Have students share the different possible ways to make figures with perimeters of 8, 12, and 14 units in Problems 4-6. Provide extra recording sheets for students to find as many figures as they can for these three perimeters involving shapes other than squares and rectangles.

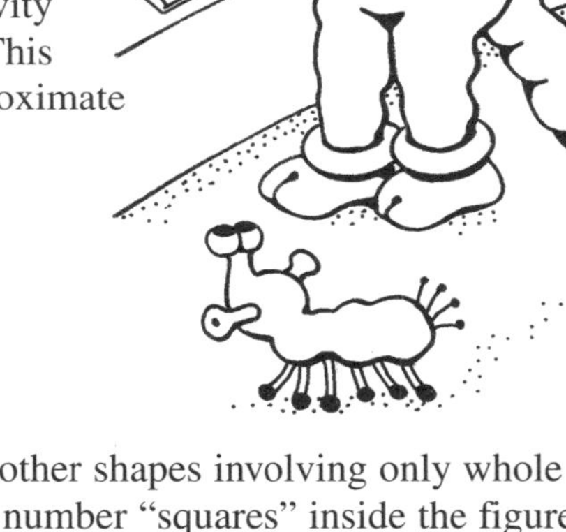

Estimate Perimeter ***(page 25)***

Only the diagonal unit of measure (/) is used in this activity although there are other diagonal units that could be used. This unit is about 1.4 units, or the square root of 2 ($\sqrt{2}$). An approximate range is given for the length of the diagonal unit, between 1¼ and 1½, to help students estimate the perimeters in Problems 1-6. By having students make their own rulers, they can measure each figure and check to see how their actual measurements compare to their estimates.

Area and Square Units ***(page 26)***

This activity deals with the area of squares, rectangles, and other shapes involving only whole square units of area. Help students understand that area is the total number "squares" inside the figure. Some students may know that the area of a rectangle or square is its length times its width ***(A = l x w).*** Have students share the different possible ways to make figures with areas of 9, 12, and 7 in Problems 4-6 making various shapes besides squares and rectangles.

Area and Halves ***(page 27)***

The area of Figure A in the example and Problems 1-3 involve whole- and half-square units of measure. However, a little different approach is needed to find the area of Figure B and solve Problems 4-6. In Figure B, count the number of whole squares and then look at the diagonal line as cutting in half a 1 x 2-unit rectangle having an area of 2 square units. One-half of that area is 1 square unit plus the other 6 whole square units is a total area of 7 square units.

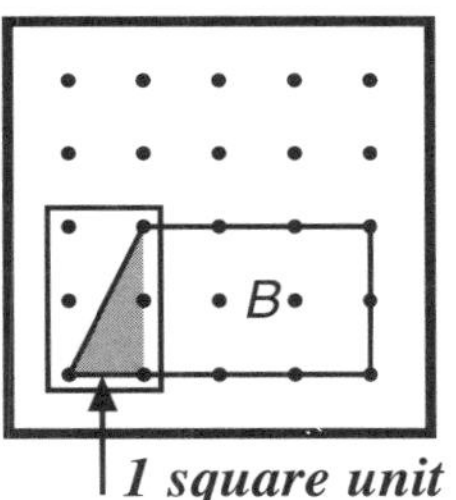

1 square unit

Area of Parallelograms and Triangles ***(page 28)***

The area formulas ***A = base x height*** and ***A = ½ x base x height*** are given for parallelograms and triangles. Students should calculate the areas and then check their answers by counting whole and half units using the technique on page 27.

Name .... Date ....

Perimeter

The *perimeter* is the distance around a figure.
It is measured in linear units.

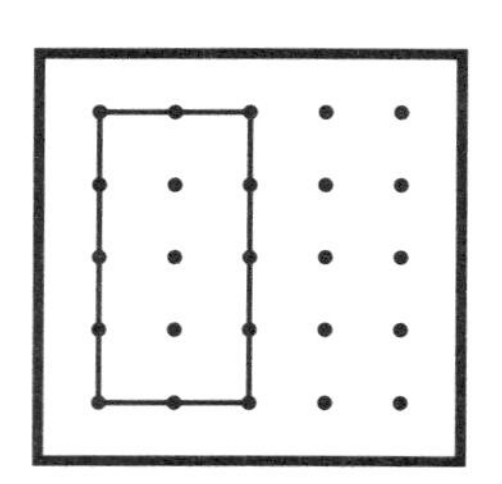

is 1 unit
The perimeter is 12 units.

Model each figure. Write the perimeter in the blank.

1. Perimeter = _____ units.

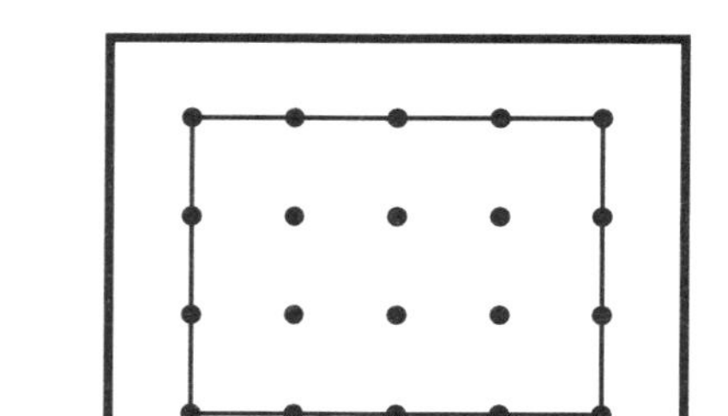

2. Perimeter = _____ units.

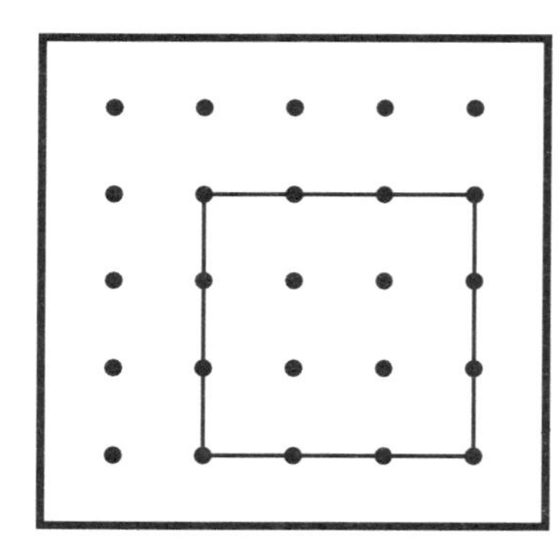

3. Perimeter = _____ units.

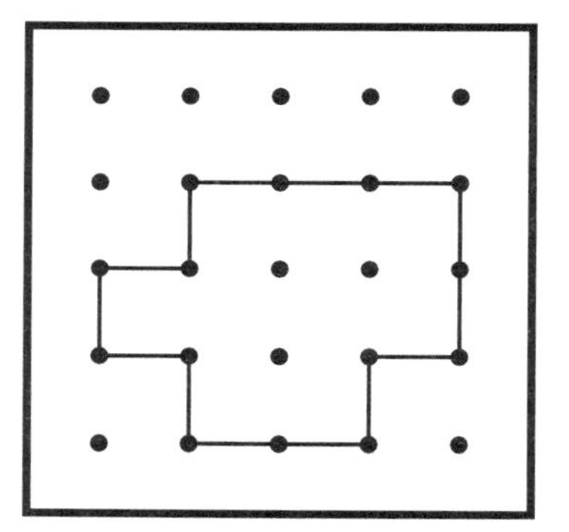

Model figures for the given perimeters. Draw them below.

4. Perimeter = 8 units.

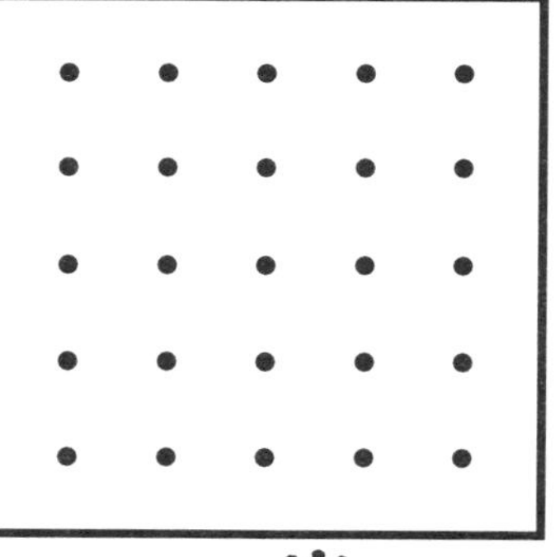

5. Perimeter = 12 units.

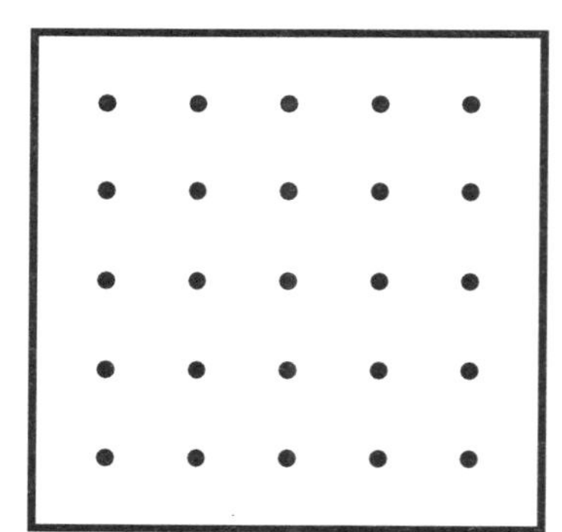

6. Perimeter = 14 units.

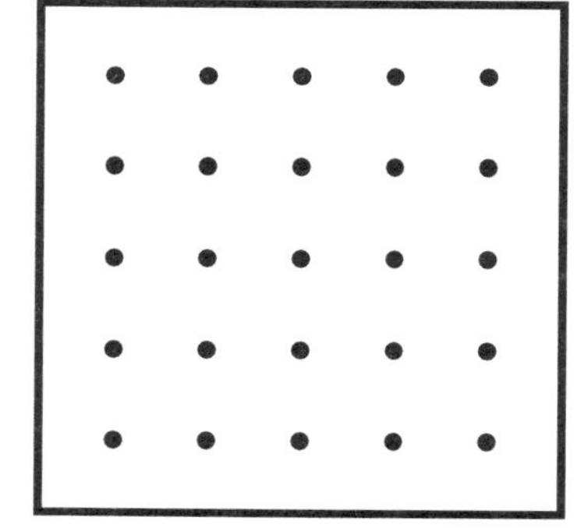

Name .... Date ....

Estimate Perimeter

When diagonal lines are used to form some or all the sides of a figure, make an estimate for the perimeter.

Since •—• is 1 unit, then ⁄ is a little more than 1 unit, about 1¼ to 1½ units.

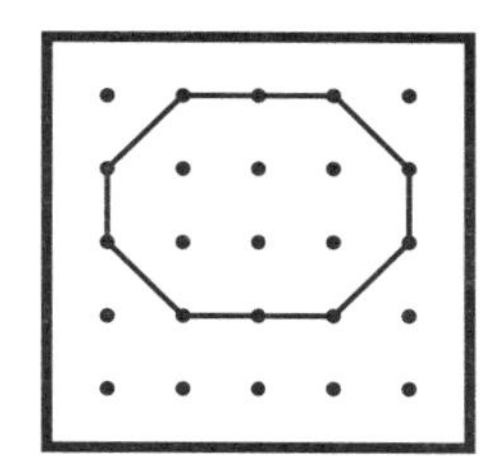

Estimated perimeter is about 12 units.

Model each figure. Estimate the perimeter.

1. Perimeter is about _____ units.

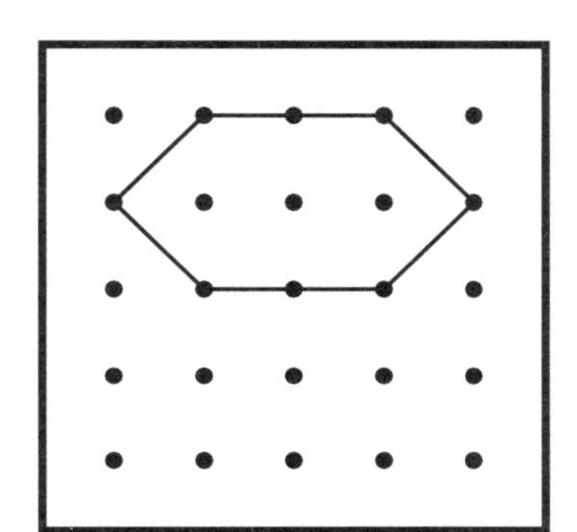

2. Perimeter is about _____ units.

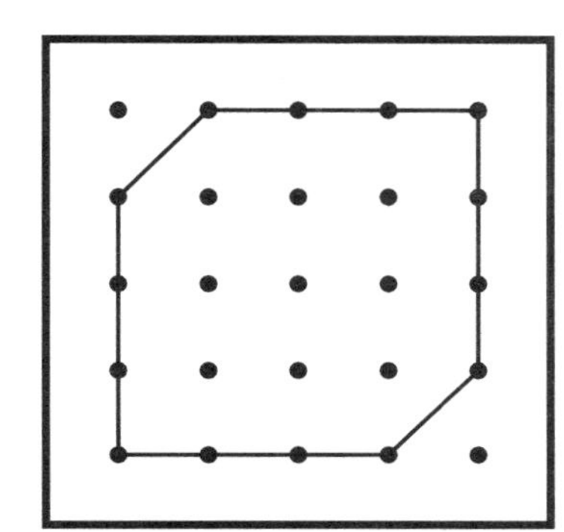

3. Perimeter is about _____ units.

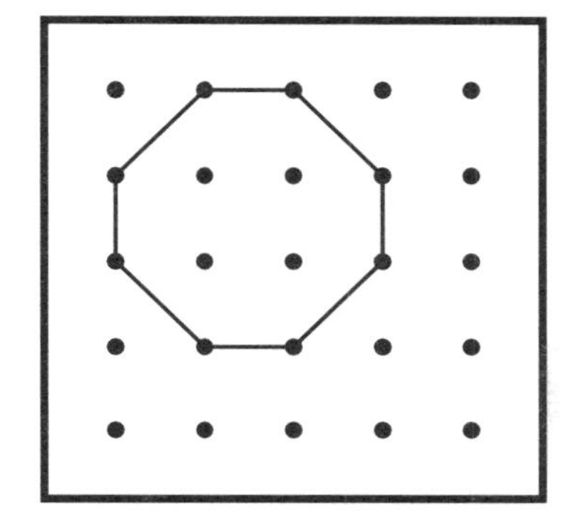

4. Perimeter is about _____ units.

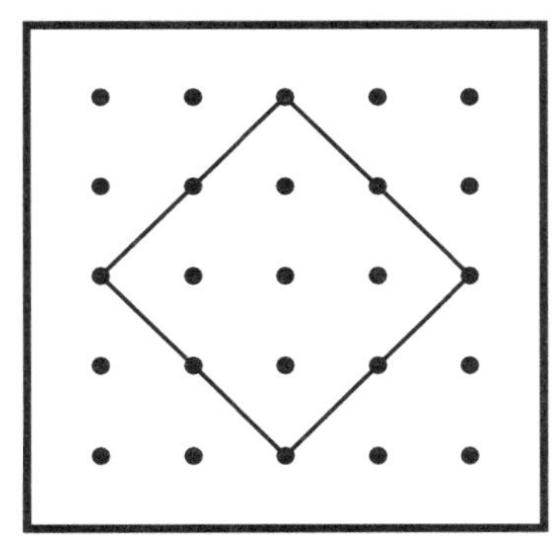

5. Perimeter is about _____ units.

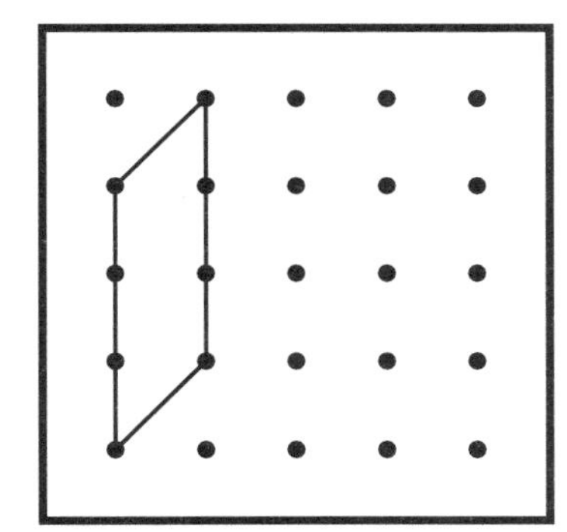

6. Perimeter is about _____ units.

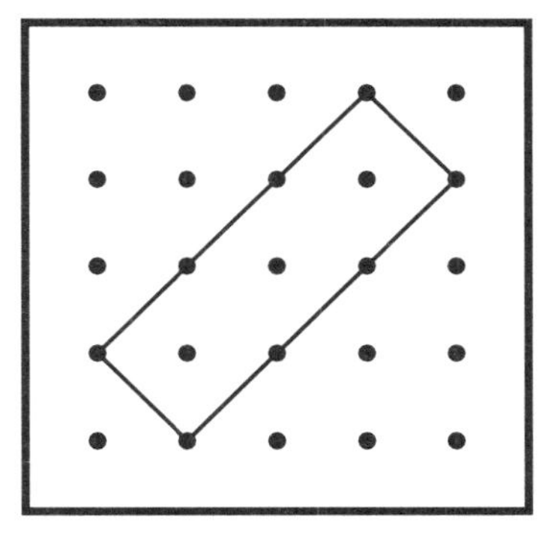

7. Make a ruler where •—• is 1 unit. (You may wish to put ½s and ¼s on your ruler too.) Measure the sides of the figures above to find the perimeters. Write your measured perimeters.

#1__________, #2__________, #3__________, #4__________, #5__________, #6__________

How close are your estimates to your actual measurements?

Name .... Date ....

Area and Square Units

Area is amount of surface covered by a figure.
It is measured in square units.

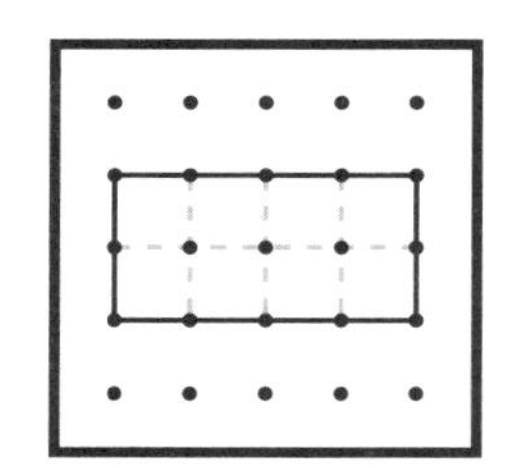

is 1 square unit
The area is 8 square units.

Model each figure. Write the area in the blank.

1. Area =
_____ square units.

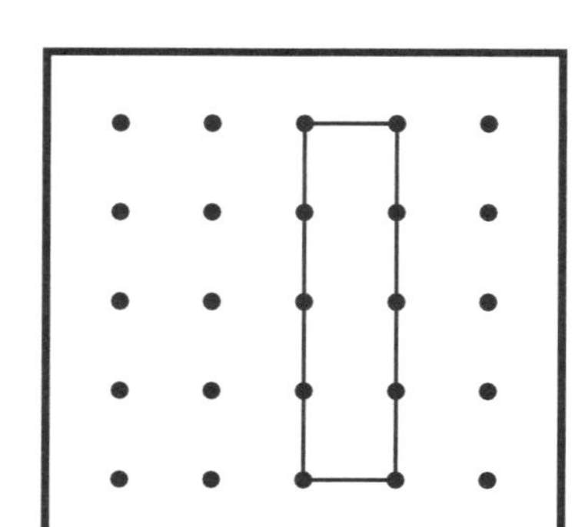

2. Area =
_____ square units.

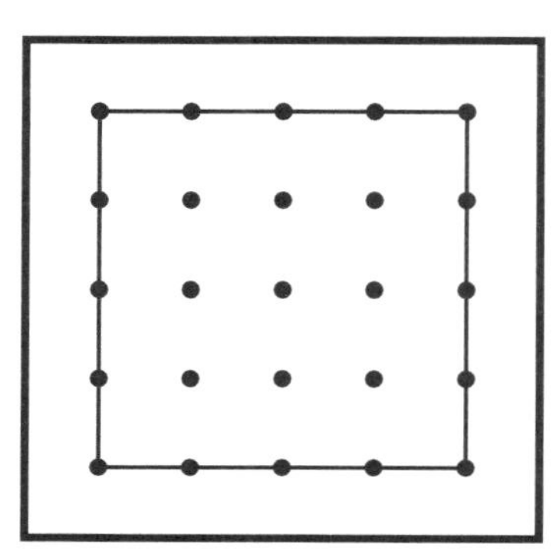

3. Area =
_____ square units.

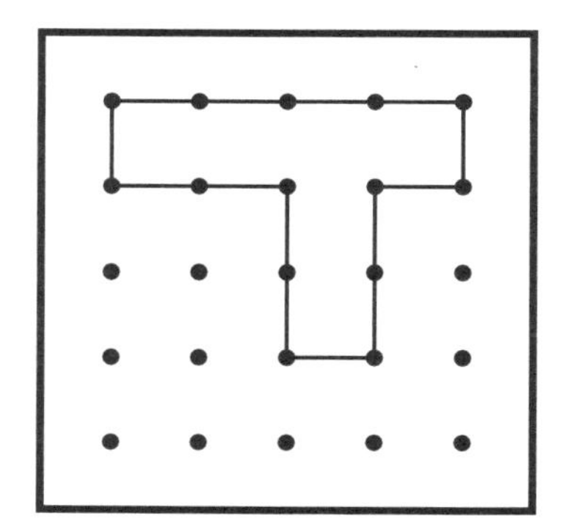

Model figures for the given areas. Draw them below.

4. Area = 9 square units.

5. Area = 12 square units.

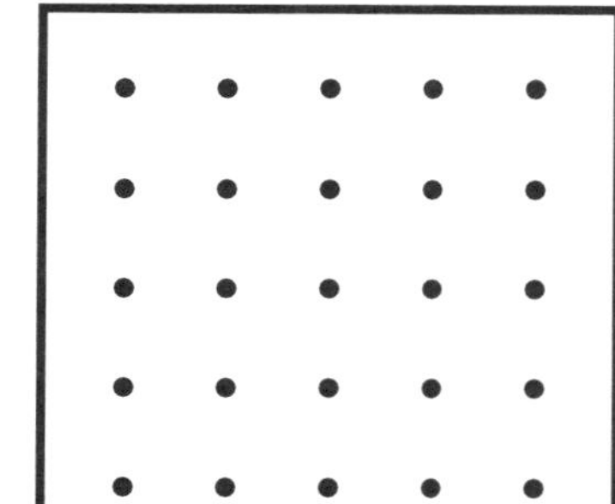

6. Area = 7 square units.

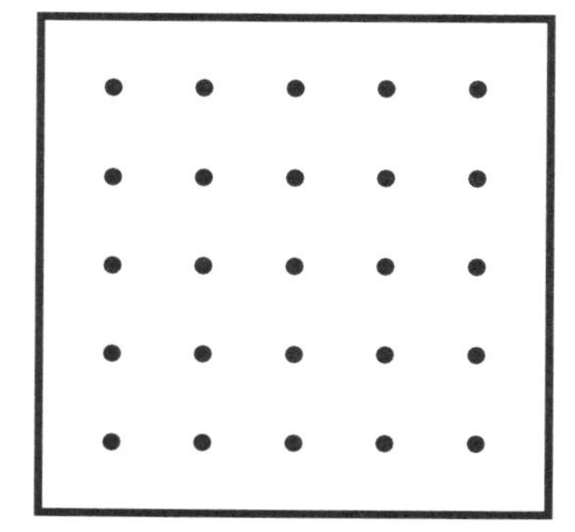

Name .... Date ....

Area and Halves

Some figures formed on the geoboard have diagonal lines for their sides.

The area of Figure *A* is 3 square units.
The area of Figure *B* is 7 square units.

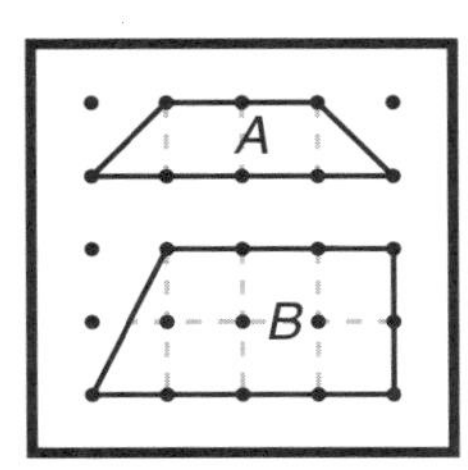

Model each figure. Write the area in the blank. (Hint: All the areas are exact, not estimates.)

1. Area = _____ square units.

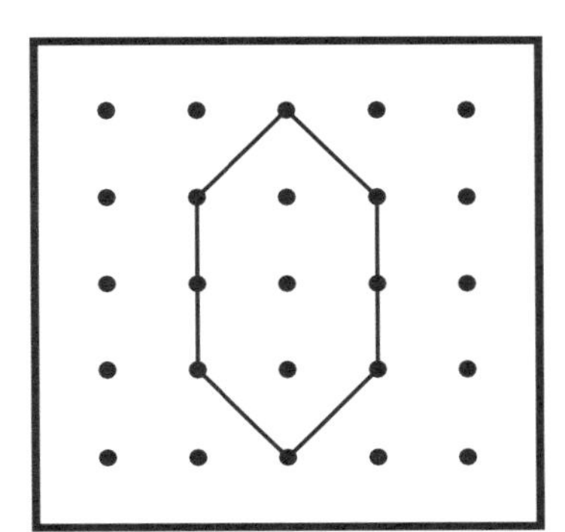

2. Area = _____ square units.

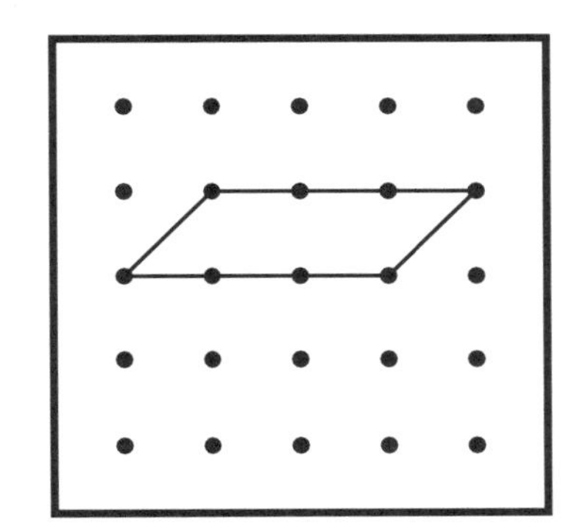

3. Area = _____ square units.

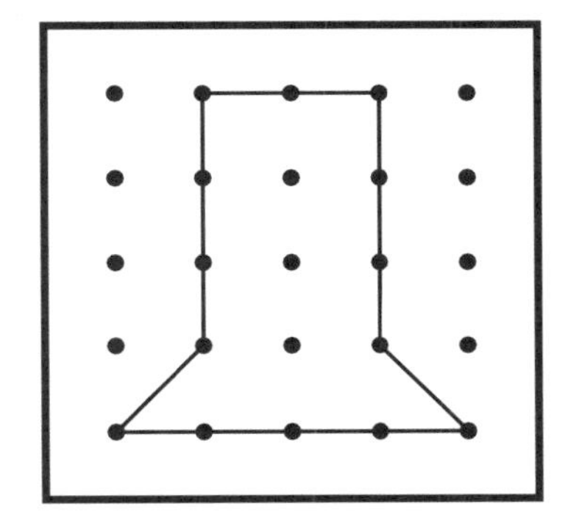

4. Area = _____ square units.

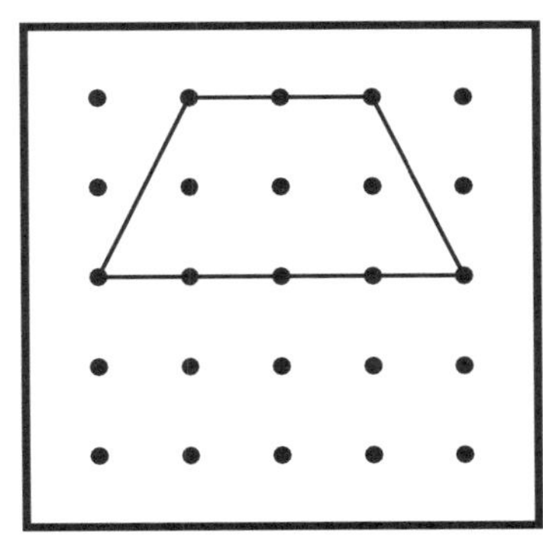

5. Area = _____ square units.

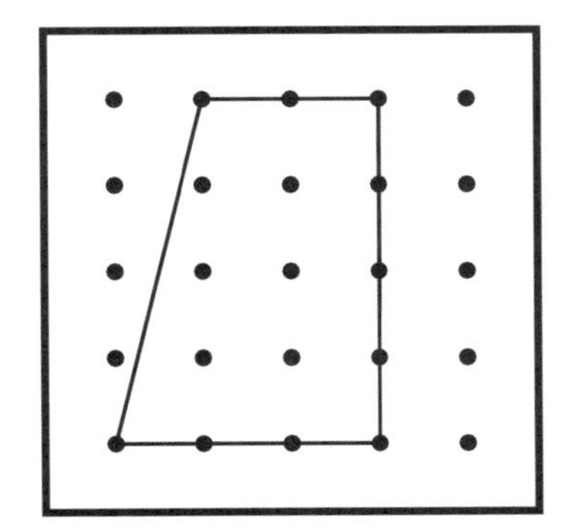

6. Area = _____ square units.

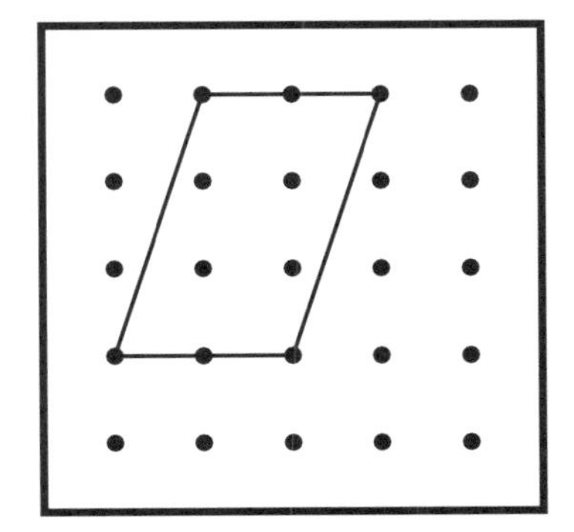

7. Explain how you found the area for the figures in Problems 4-6.

____

____

____

Name .... Date ....

Area of Parallelograms and Triangles

Since triangles and parallelograms have sides with diagonal lines, use whole- and half-square units to find area. You also can use these area formulas:

Parallelogram: Area = base x height
Triangle: Area = ½ x base x height

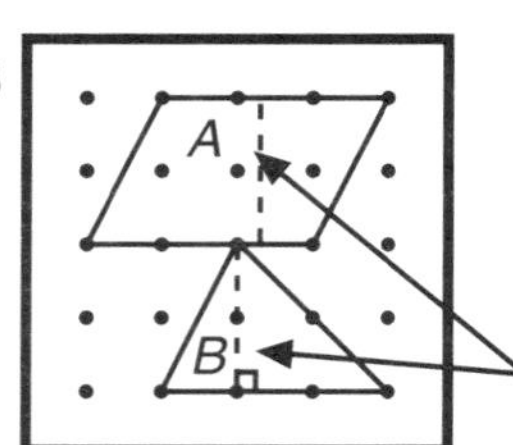

Figure A:
A = 3 x 2 = 6 square units
Figure B:
A = ½ x 3 x 2 = 3 square units.
height

Model each of the parallelograms and triangles. Calculate the area and write it in the blank.

1. Area = _____ x _____

= _____ square units.

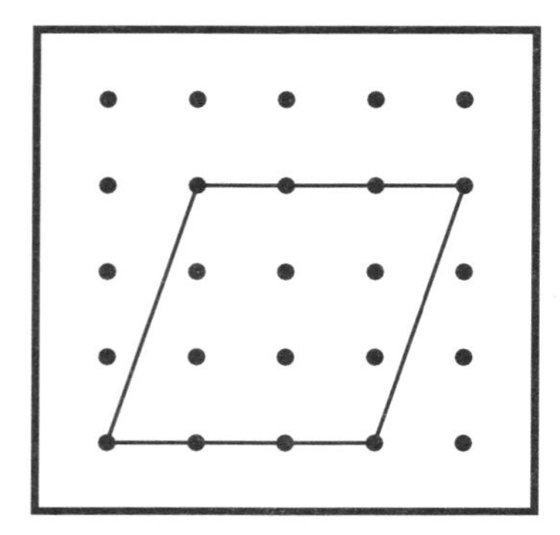

2. Area = _____ x _____

= _____ square units.

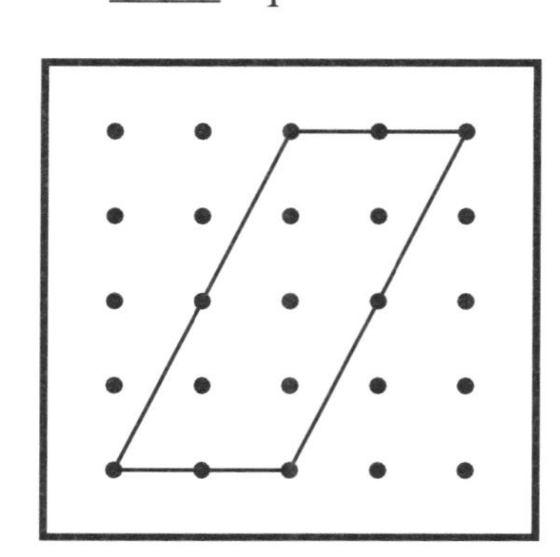

3. Area = _____ x _____

= _____ square units.

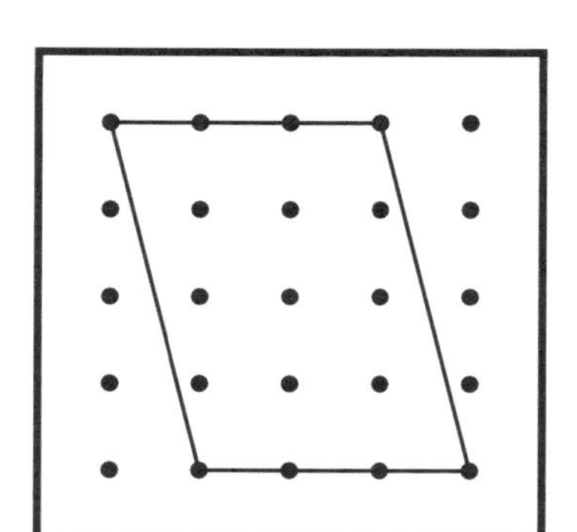

4. Area = ½ x _____ x _____

= _____ square units.

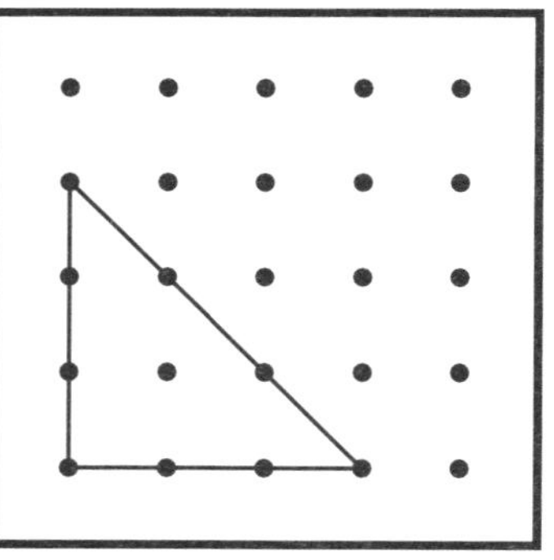

5. Area = ½ x _____ x _____

= _____ square units.

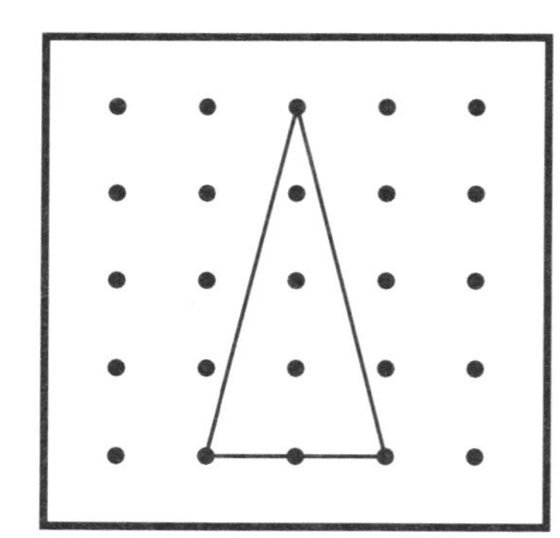

6. Area = ½ x _____ x _____

= _____ square units.

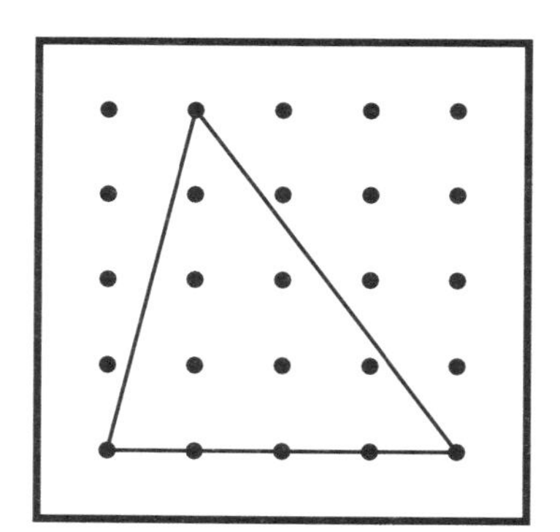

Fractions

Vocabulary

equivalent fractions, fraction, proper fraction, unit fraction

Activity Teaching Notes

Halves, Fourths, and Eighths ***(page 30)***
When one rubber band is placed around all the outer pins on the geoboard, this square shape represents the number 1 (1 whole). Students can use more rubber bands to divide the square into 2, 4, and 8 equal parts to show $\frac{1}{2}$s, $\frac{1}{4}$s, and $\frac{1}{8}$s. Tell students to focus on one of the parts to see how its size compares to the entire geoboard square. The unit fraction $\frac{1}{2}$, $\frac{1}{4}$, or $\frac{1}{8}$ represents one part of the entire square region. Extend the activity by providing students with recording sheets and asking them to show $\frac{1}{16}$s and $\frac{1}{32}$s as many ways as they can.

Dividing the large square on the geoboard is similar to working with fraction square models. Using 4 x 4-inch squares of paper, students can fold and cut the same fraction parts from paper as they model them on the geoboard.

Model Equal Parts ***(page 31)***
This is another activity to show equal-size parts of a whole region and focus on unit fractions. Unlike the activity on page 30 where all the whole regions were squares, the regions for this activity involve a variety of geometric shapes. As students discuss how they divided each region into equal-size parts, also review the geometric names of each region.

Given One Part, Show the Whole ***(page 32)***
In this activity, students again work with unit fractions after being shown one part of a whole region. Encourage them to model a geometric shape for the solution to each problem. Discuss each problem by asking students to given a fraction for the part needed to complete the whole region. This activity provides practice in visual reasoning as well as readiness for addition of fractions (i.e., $\frac{1}{3}$ and $\frac{2}{3}$ is 1).

Given Parts, Show the Whole ***(page 33)***
In this activity, students are given a proper fraction and must visualize what the whole region looks like. After they find the solutions, discuss each problem and ask what part of the whole region added on to the given part to make it a whole region. This activity will help students prepare for addition of fractions.

Fraction Parts, Sums of One ***(page 34)***
There are two ways that students can model the addition problems. They can use one rubber band for each addend, or they can use one rubber band for one addend and the other rubber band to show the whole region. Have students share all the different possible ways to model these sums.

Name .... Date ....

Halves, Fourths, and Eighths

Put one rubber band around all the outer pins to show a large square.
Use other rubber bands to show equal parts. Record your solutions below.

1. Show 2 equal parts three different ways. Each part is ½ .

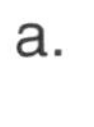

a. 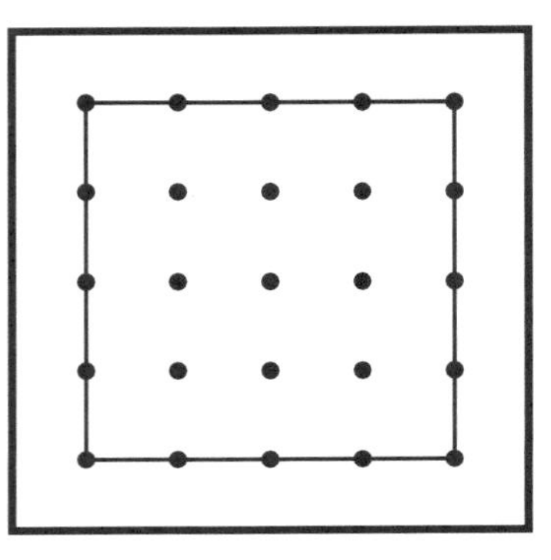 b. 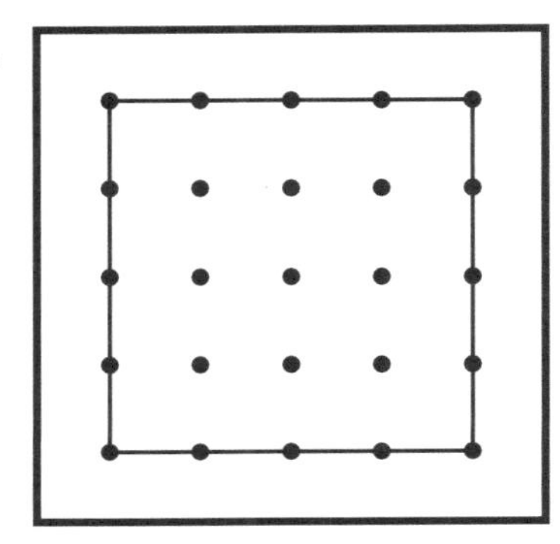 c. 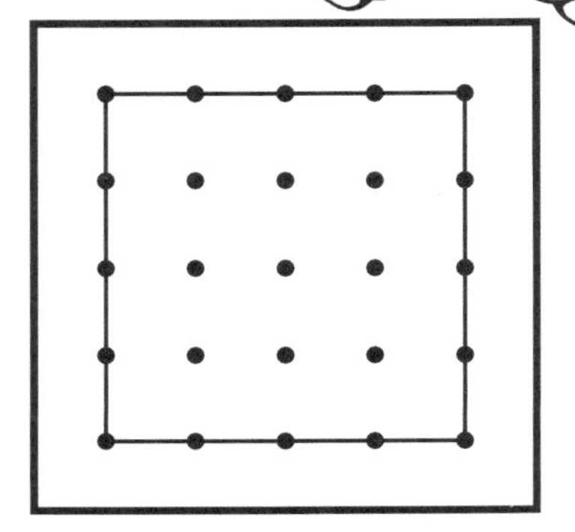

2. Show 4 equal parts three different ways. Each part is ______.

a. 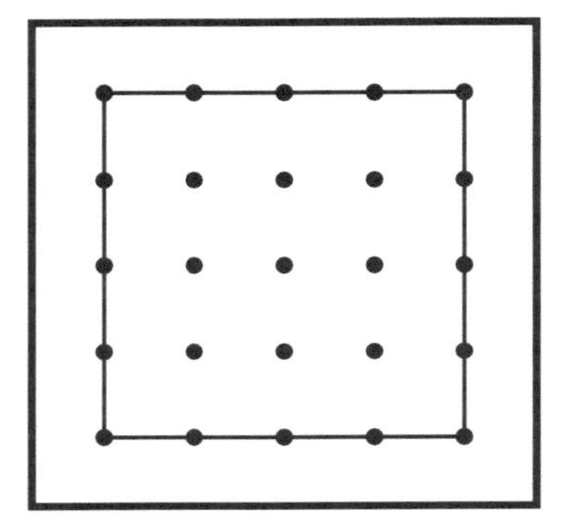 b. 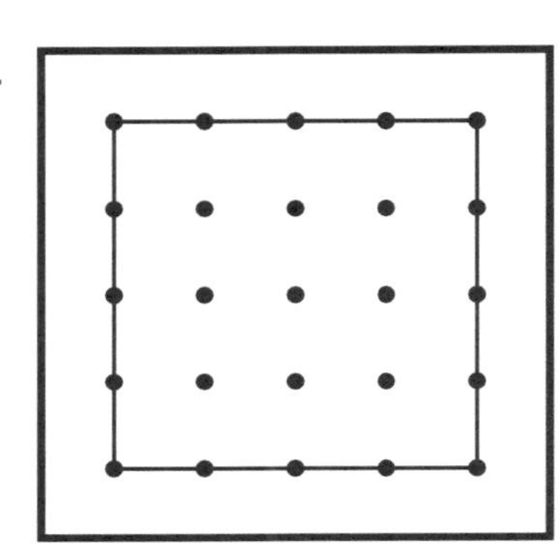 c. 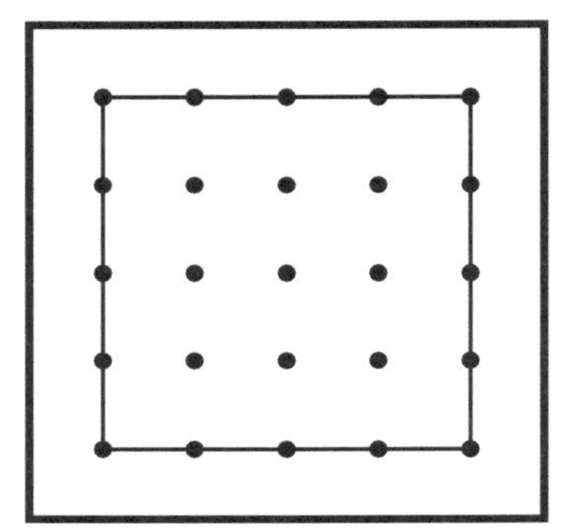

3. Show 8 equal parts three different ways. Each part is _____.

a. 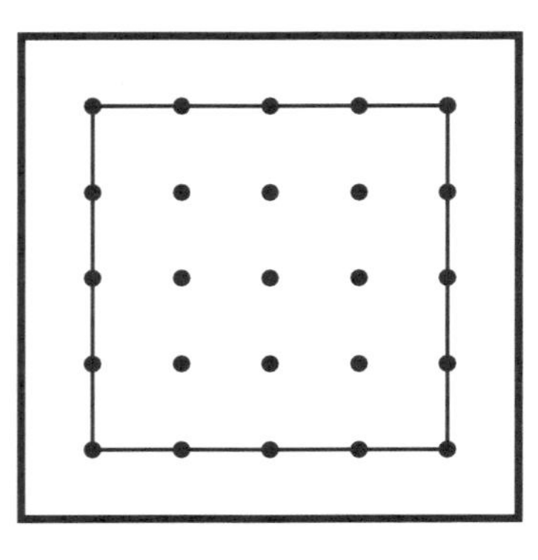 b. 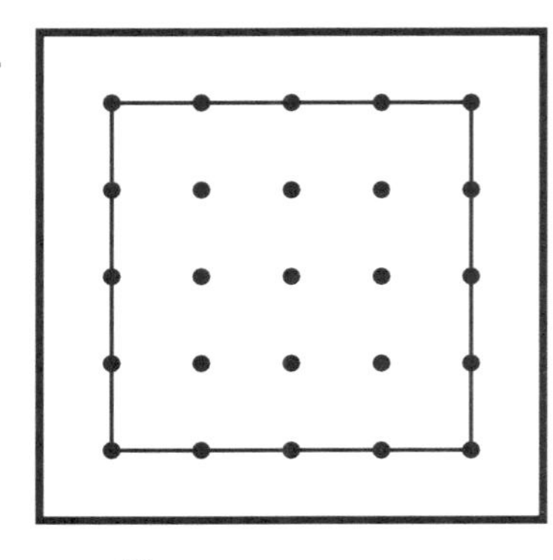 c. 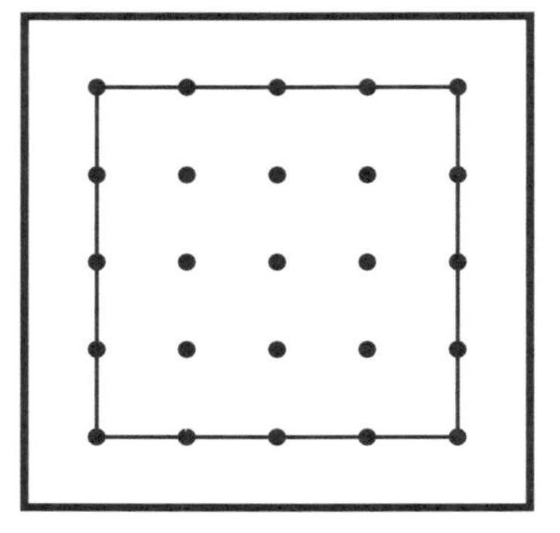

Name .... Date ....

Model Equal Parts

Use one rubber band to model each figure. Use other rubber bands to show equal parts. Record your solutions below.

1. Show eighths

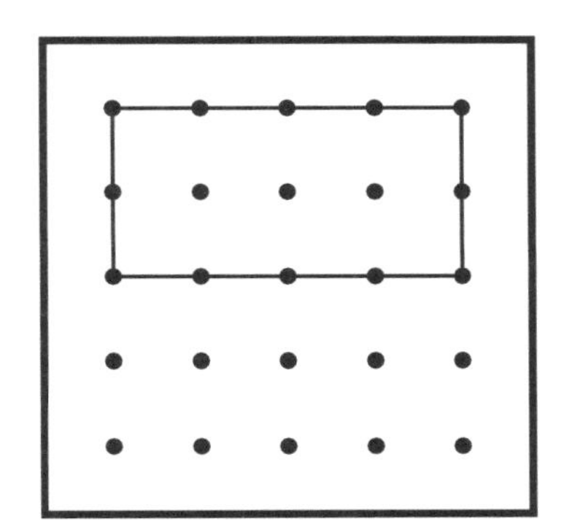

2. Show sixths

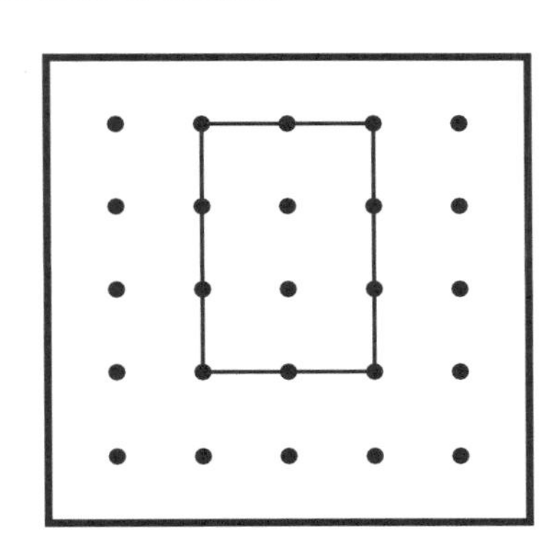

3. Show fourths

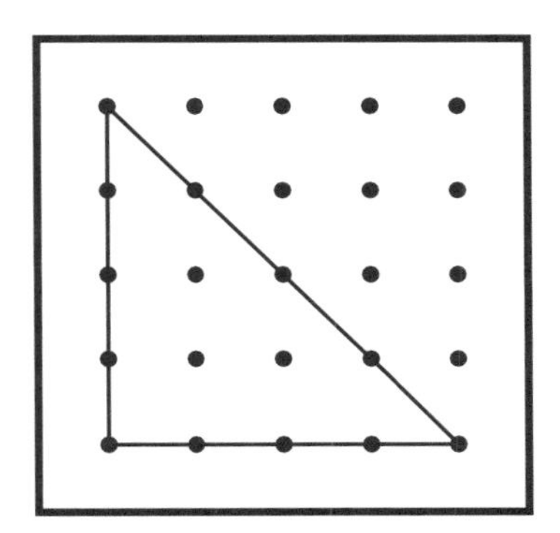

4. Show thirds

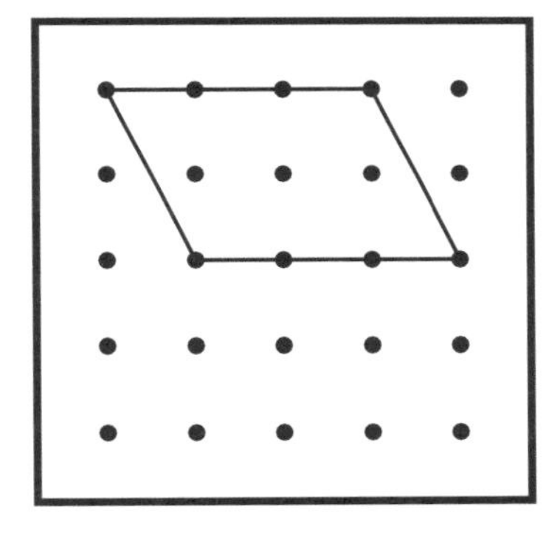

5. Show halves

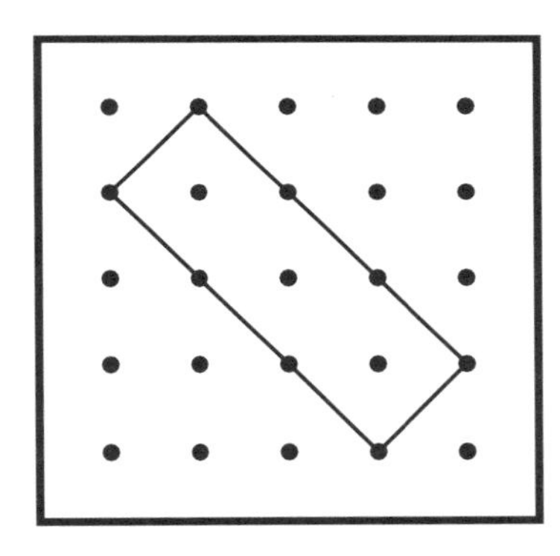

6. Show eighths

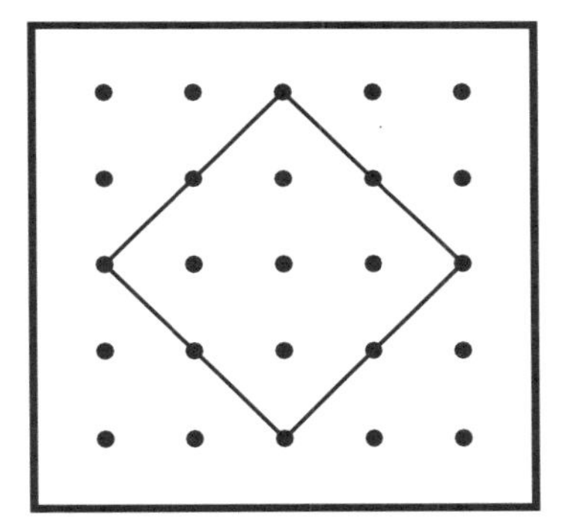

7. Show fourths

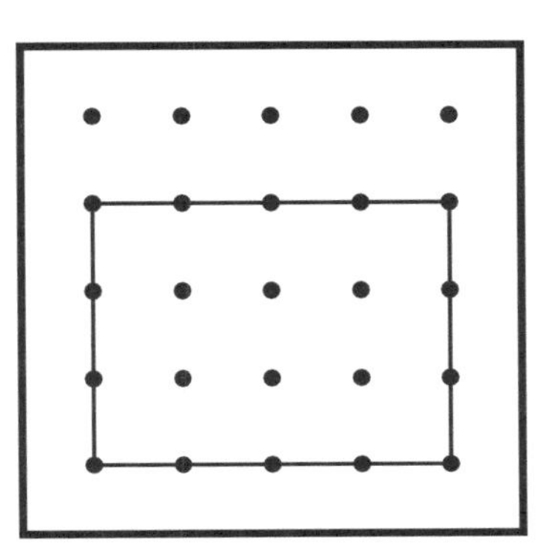

8. Show sixths

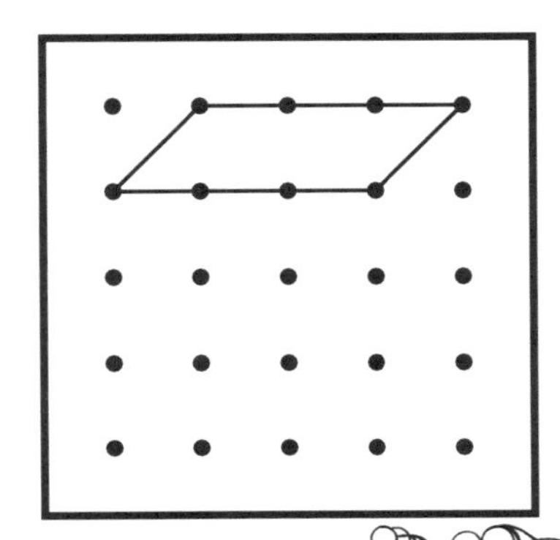

9. Show eighths

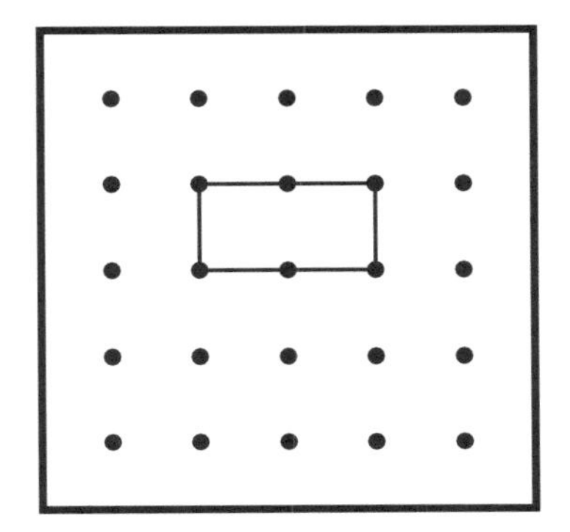

Name .... Date ....

Given One Part, Show the Whole

One part of a region is shown. Model the whole region as a geometric shape if possible. Record your solutions below.

1. This is ½.

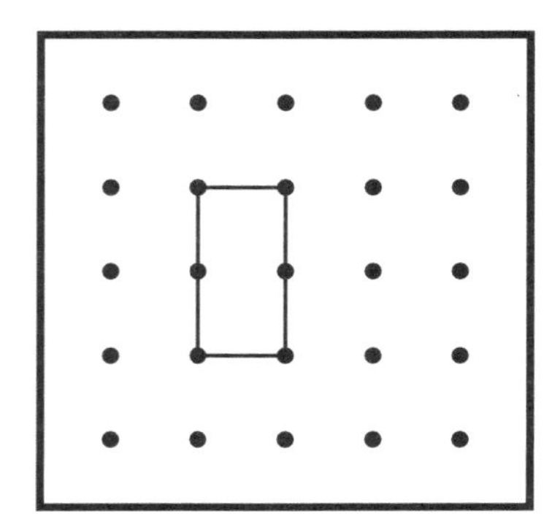

2. This is ⅓.

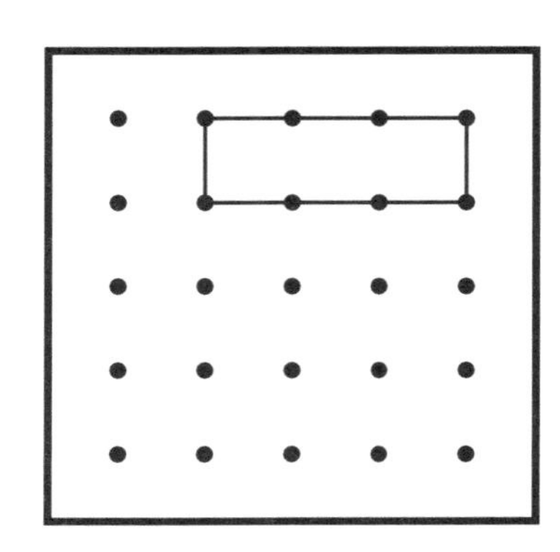

3. This is ¼.

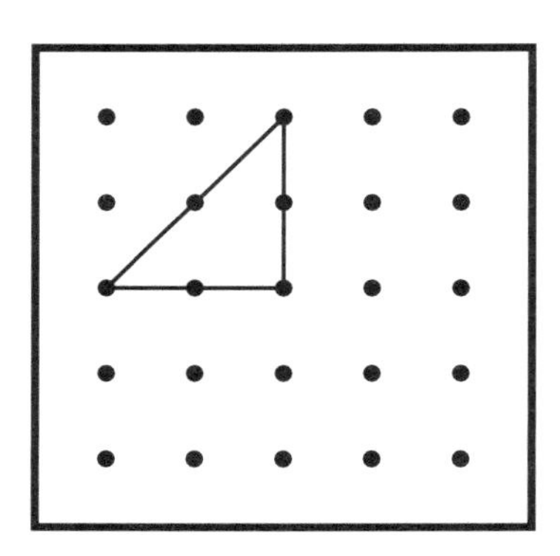

4. This is ⅕.

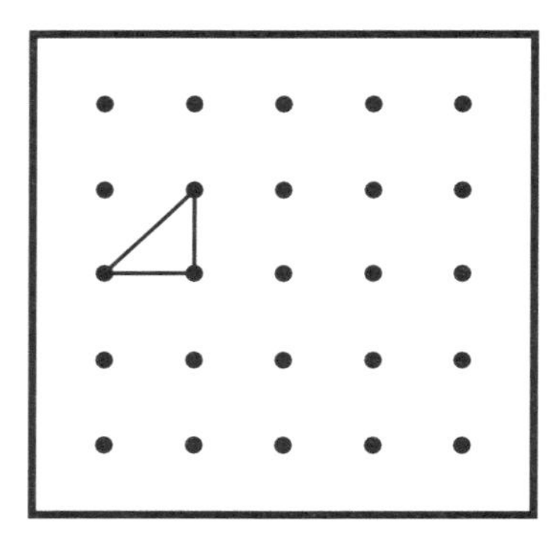

5. This is ⅙.

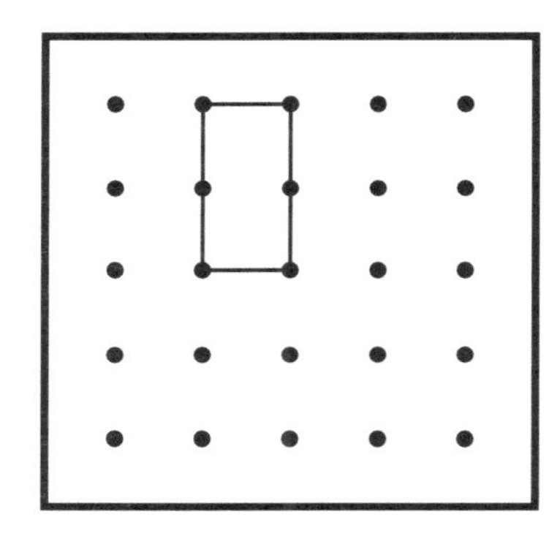

6. This is ⅐.

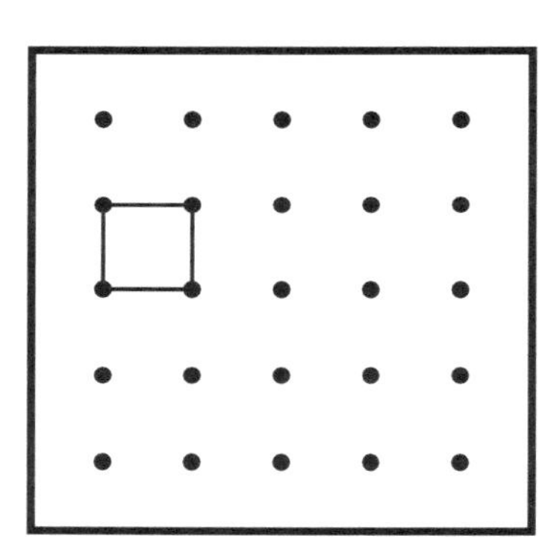

7. This is ⅛.

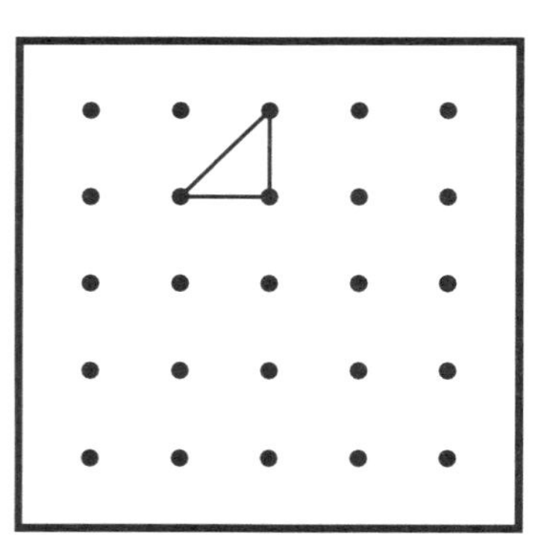

8. This is ⅒.

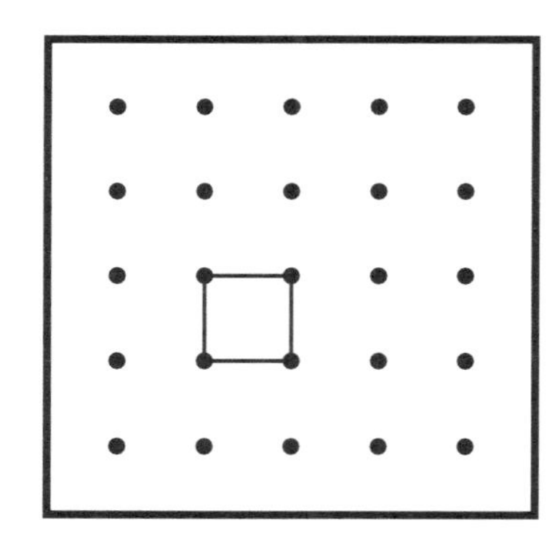

9. This is 1/12.

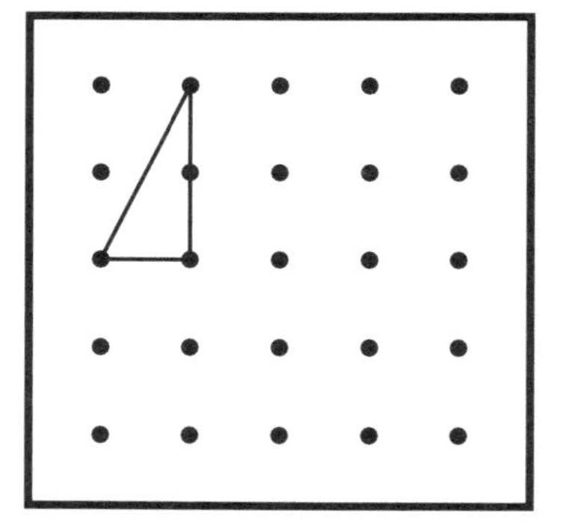

Name .... Date ....

Given Parts, Show the Whole

Several parts of a region are shown. Model the whole region.
Use one rubber band to model the part shown. Use another rubber band or bands to model the whole region. Record the whole regions below.

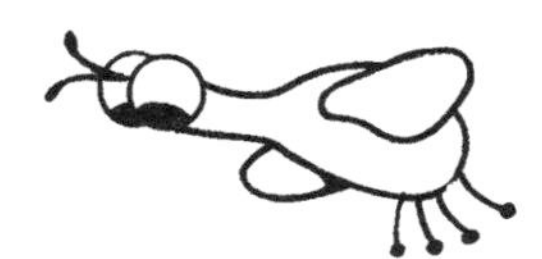

1. This is $\frac{2}{3}$.

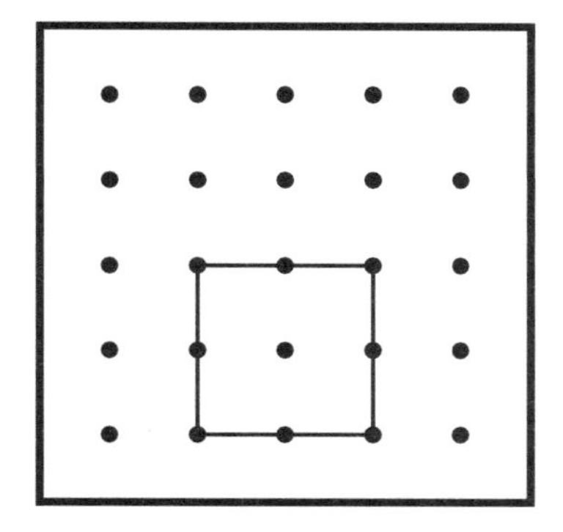

2. This is $\frac{3}{4}$.

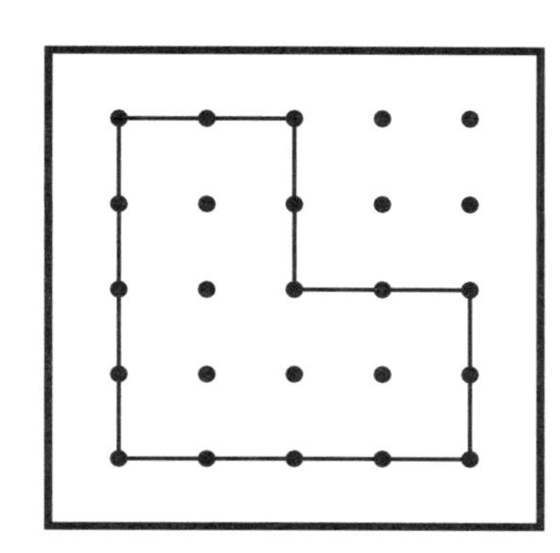

3. This is $\frac{5}{8}$.

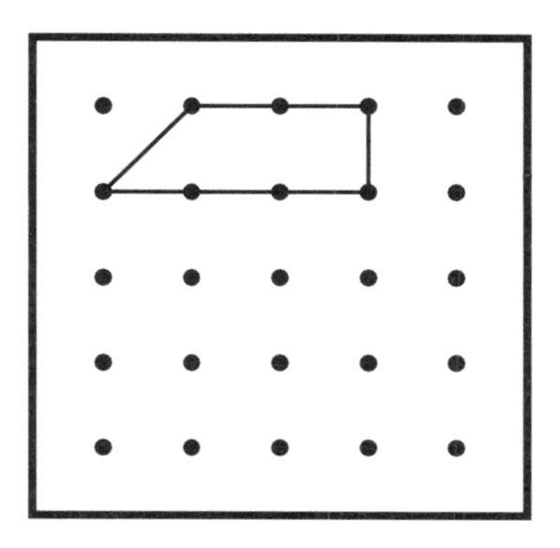

4. This is $\frac{3}{5}$.

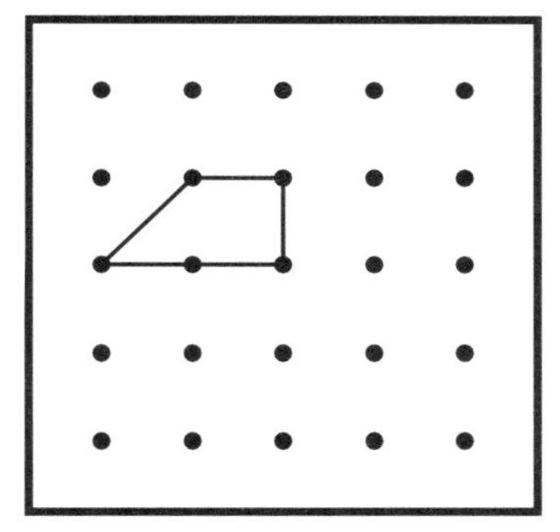

5. This is $\frac{5}{6}$.

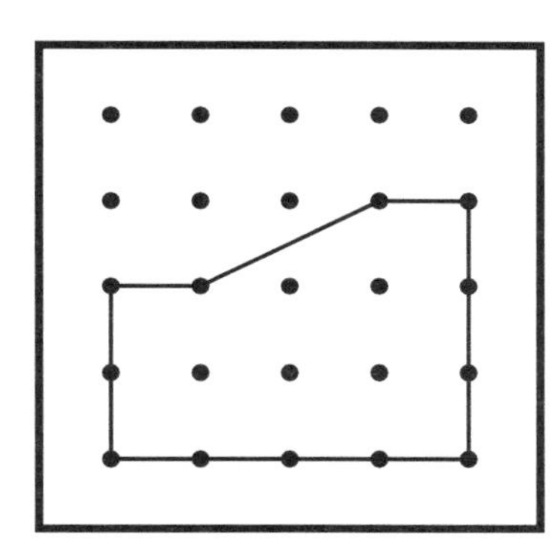

6. This is $\frac{3}{4}$.

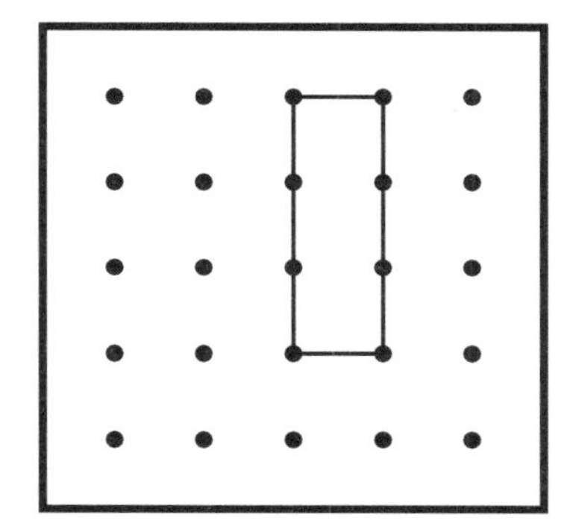

7. This is $\frac{3}{8}$.

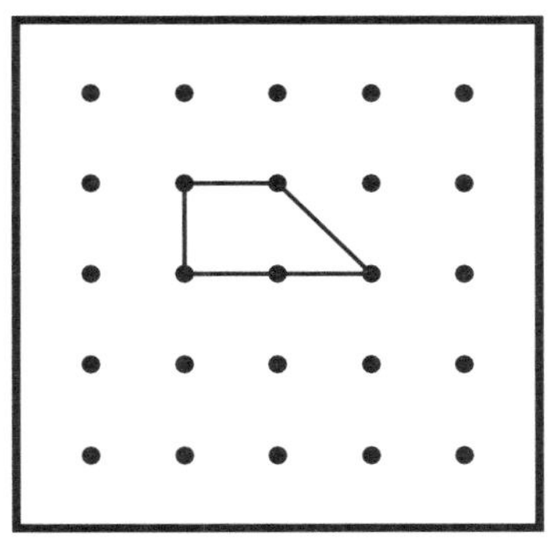

8. This is $\frac{4}{5}$.

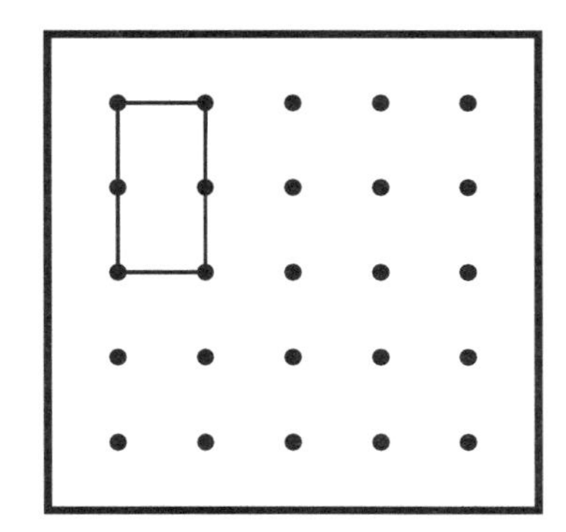

9. This is $\frac{5}{12}$.

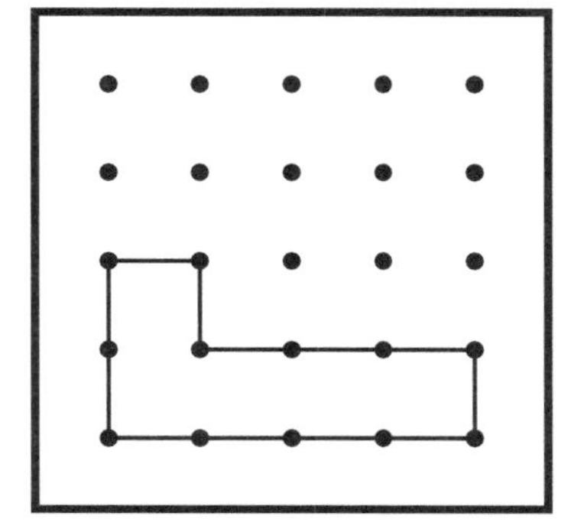

Name .... Date ....

Fraction Parts, Sums of One

Use two rubber bands to model the addition problems. Record them below. The first problem is done for you.

1. $\frac{2}{3} + \frac{1}{3} = 1$

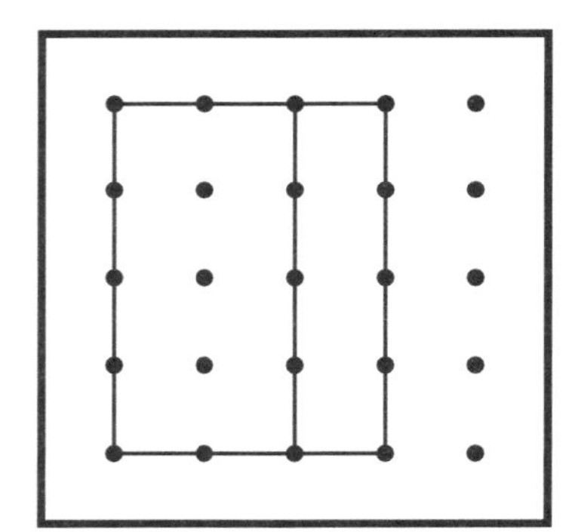

2. $\frac{3}{4} + \frac{1}{4} = 1$

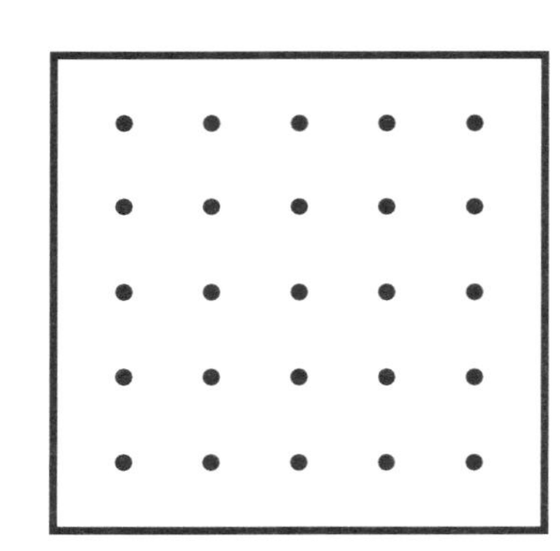

3. $\frac{1}{6} + \frac{5}{6} = 1$

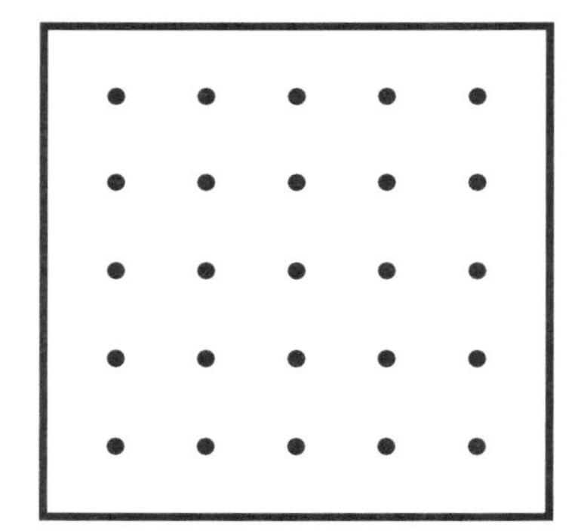

4. $\frac{3}{8} + \frac{5}{8} = 1$

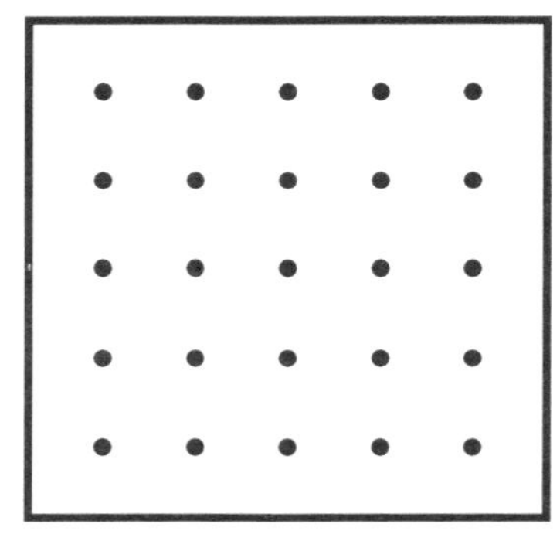

5. $\frac{2}{9} + \frac{7}{9} = 1$

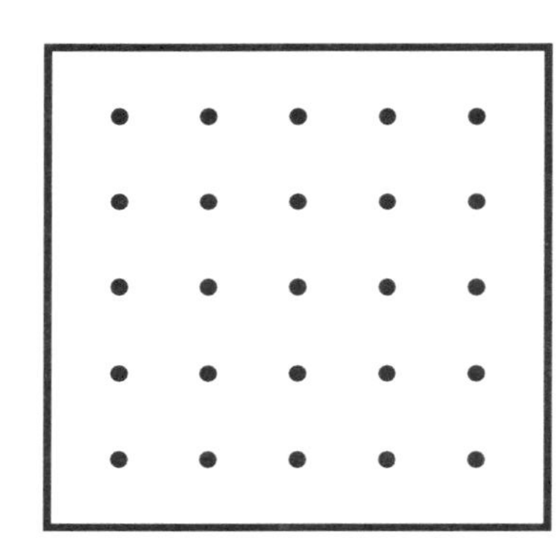

6. $\frac{1}{12} + \frac{11}{12} = 1$

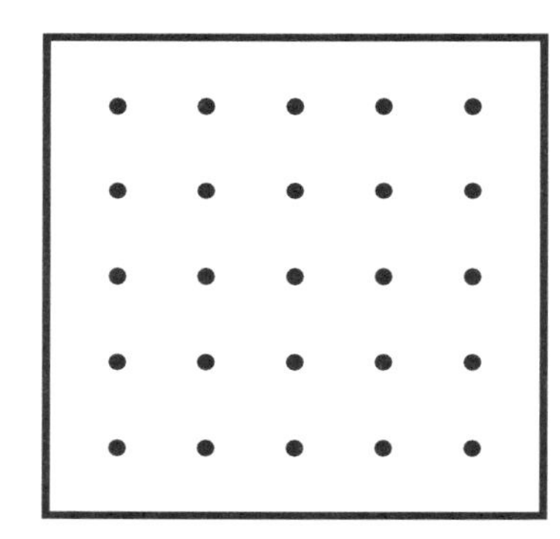

7. $\frac{5}{12} + \frac{7}{12} = 1$

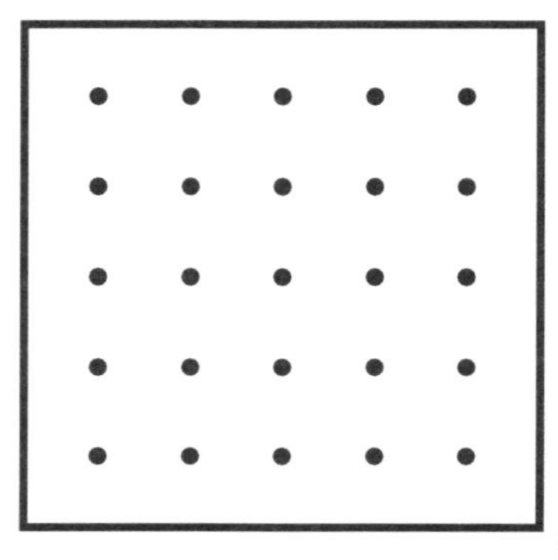

8. $\frac{4}{16} + \frac{12}{16} = 1$

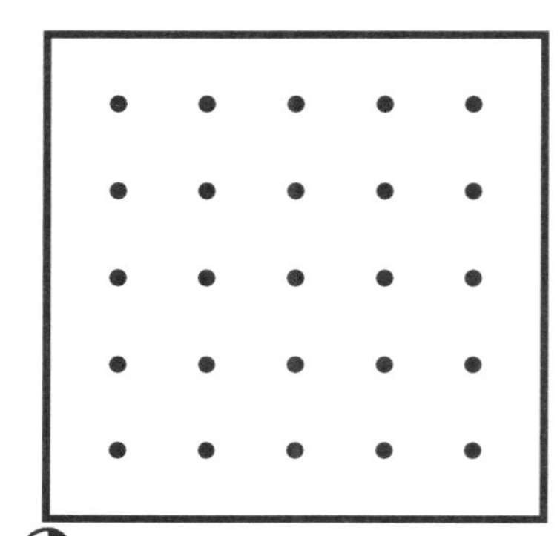

9. $\frac{3}{10} + \frac{7}{10} = 1$

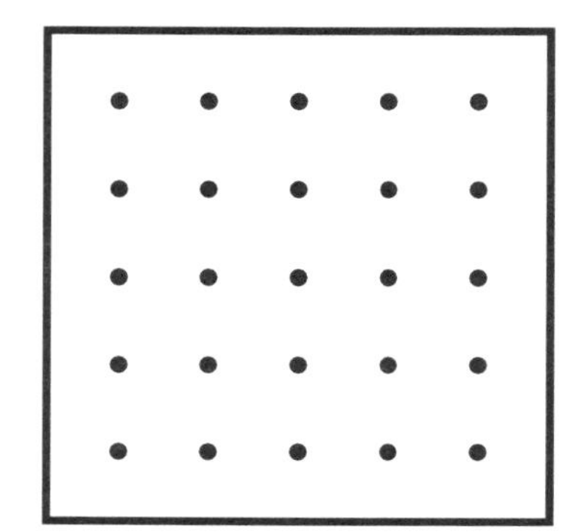

Parts of a Circle

Vocabulary

arc, central angle, chord, circle, circumference, diameter, radius (radii), vertex

Activity Teaching Notes

Circle and Radius ***(page 36)***
Students model different radii on their geoboards to discover that they are all the same length for a circle. Ask them whether there would be more radii for their circle if there were more pins on their geoboards. Since there are an infinite number of points on a circle, there would be infinite number of radii. With one or more radii shown on their geoboards, have them turn the geoboards and talk about the length and position of them.

Chords and Diameter ***(page 37)***
Students only can show six different chords on their 12-pin circular geoboard, the longest of which is the diameter of the circle. Discuss the answers to Problems 2-4. Continue the discussion by asking students about how many different diameters they could show on their geoboard and how many diameters they think any circle contains. Ask students how a radius and a diameter are related. A diameter is twice as long as a radius.

Central Angles ***(page 38)***
In this activity, students form different-sized central angles. They also form different-sized arcs of the circle. One of the central angles is a straight line, and cuts off an arc that goes half way around the circle which is a semicircle. Central angles can be larger than the central angle that forms a semicircle.

Circumference and Diameter ***(page 39)***
This activity shows students how to discover the relationship between the circumference of a circle and its diameter. The ratio of the circumference to the diameter is approximately 3.14 to 1. Allow students to use calculators to do the division. Ask them to show their answers to the nearest hundredth. After students measure at least three round objects, discuss the relationship between the circumference and radius of a circle. For a class project, tell them to find the average of all their decimal answers. Does it come close to 3.14?

Name .... Date ....

Circle and Radius

A ***circle*** is a simple, closed curve in a plane. Every point on the circle is an equal distance from the ***center.***

A ***radius*** is the distance from the center of the circle to a point on the circle.

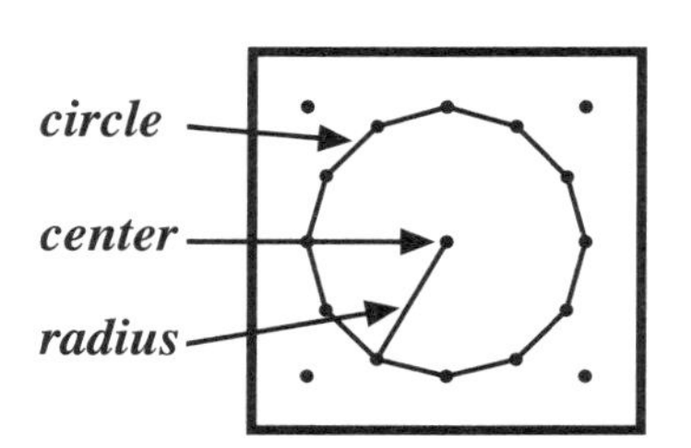

1. Place one rubber band around all the pins that form a circle. Record it at the right.

2. Model a radius of the circle with one rubber band. Record it at the right.

3. Model another radius with one more rubber band. Record it at the right.

4. Put one rubber band around the pins that form a circle. Model many different ***radii*** (plural for radius) on your geoboard. Record them below.

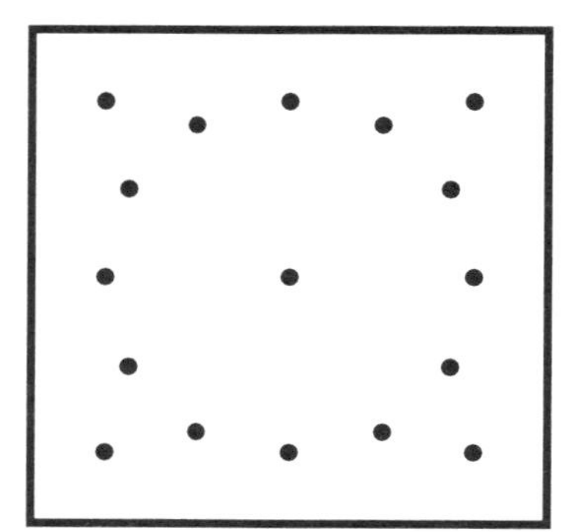

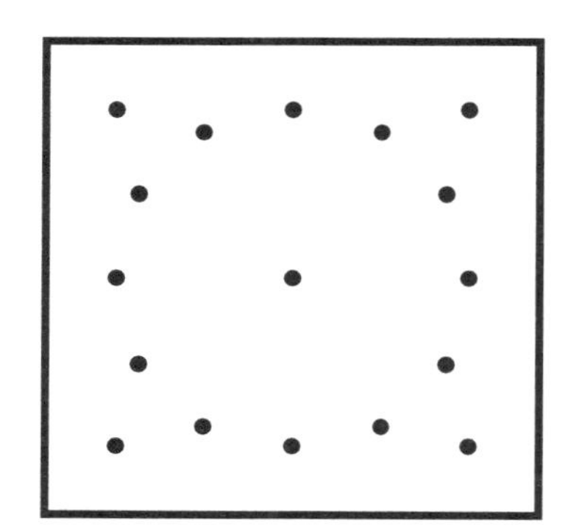

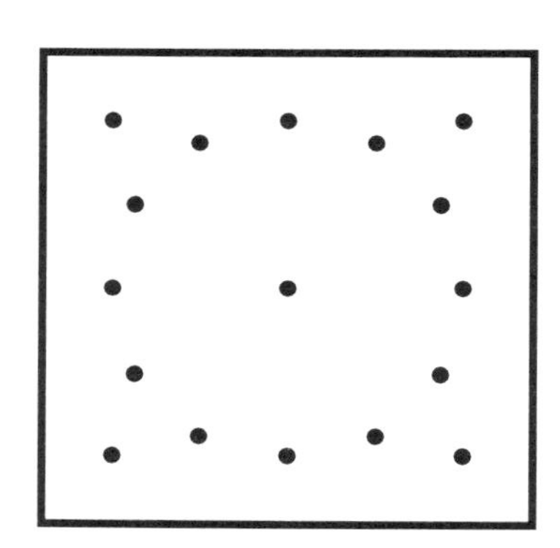

5. Do you think all the ***radii*** are the same distance from the center to a point on the circle? Why or why not?

____

____

____

____

Name .... Date ....

Chords and Diameter

A ***chord*** is a line segment connecting two points on a circle.

A ***diameter*** is a chord that passes through the center of the circle.

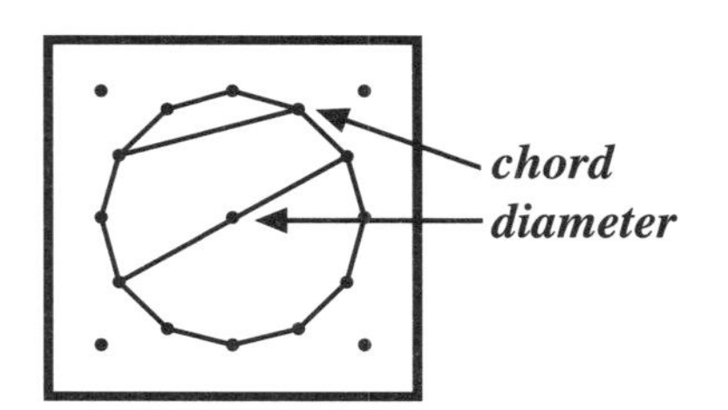

1. Place one rubber band around all the pins that form a circle.
Model six different chords on your geoboard. Record them below.

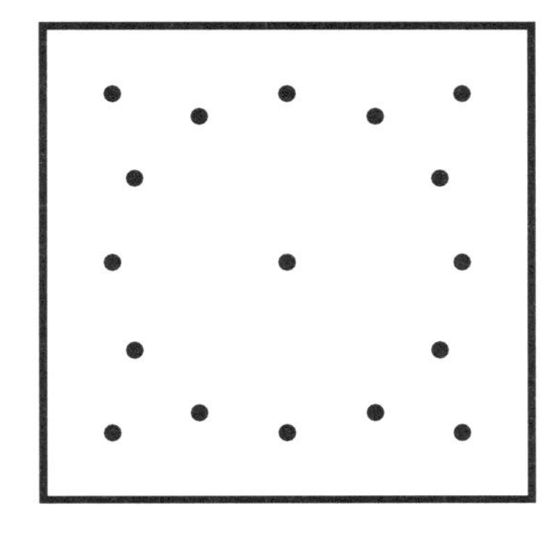
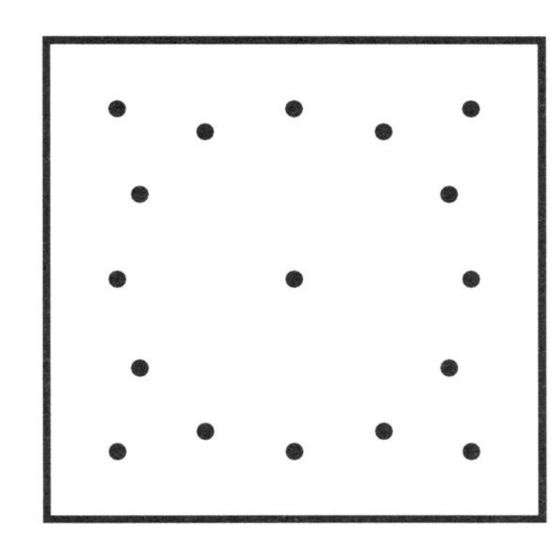
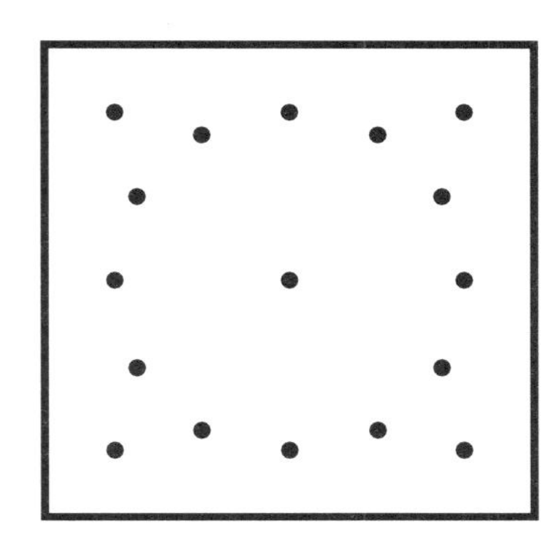

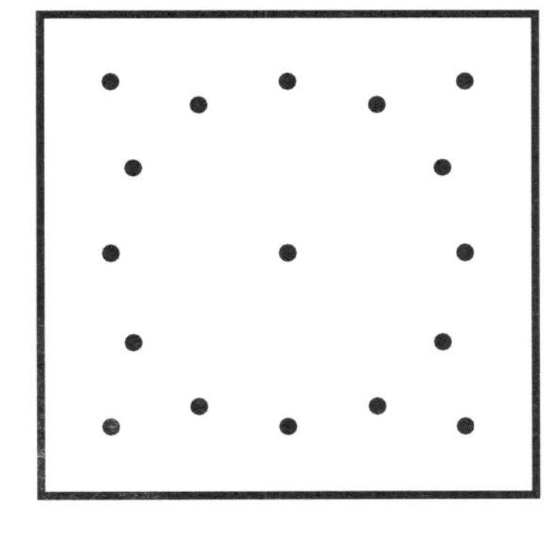
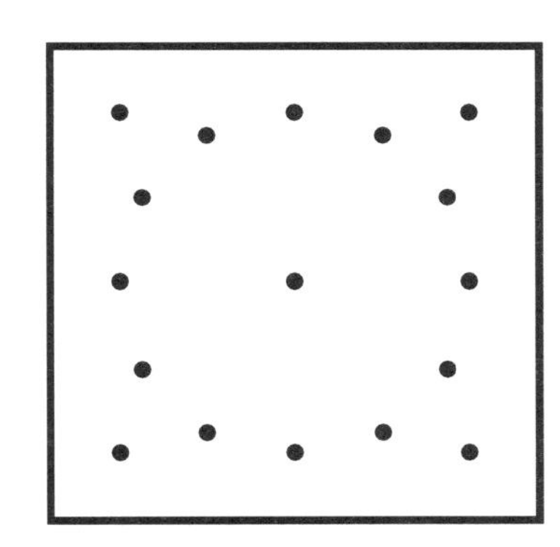
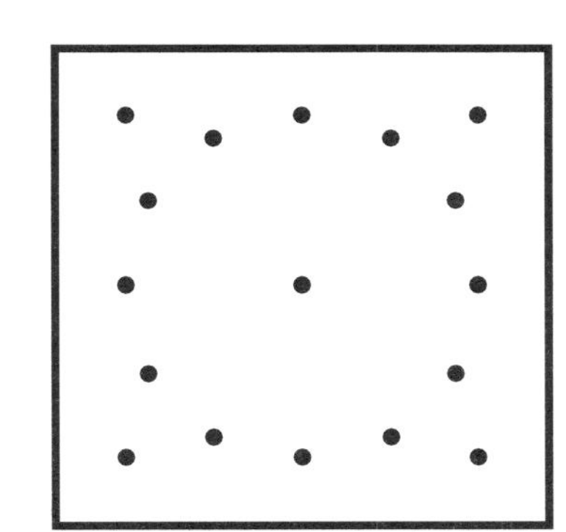

2. Did any of the chords above pass through the center of the circle? __________
If so, circle it.

3. How many different chords are on your geoboard? __________

4. What is the longest chord of the circle? ______________________
Why? ____
____
____
____

Name .... Date ....

Central Angles

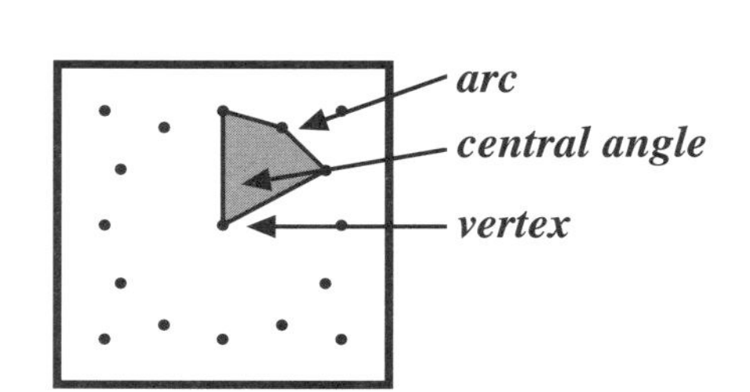

A ***central angle*** is an angle formed by two radii of a circle. The vertex is the center of the circle.

The ***vertex*** is the common endpoint of two sides that form the central angle.

An ***arc*** is part of a circle cut off by the sides of a central angle.

Use one rubber band to model five different central angles with arcs, one at a time, on your geoboard. Record them below.

1.

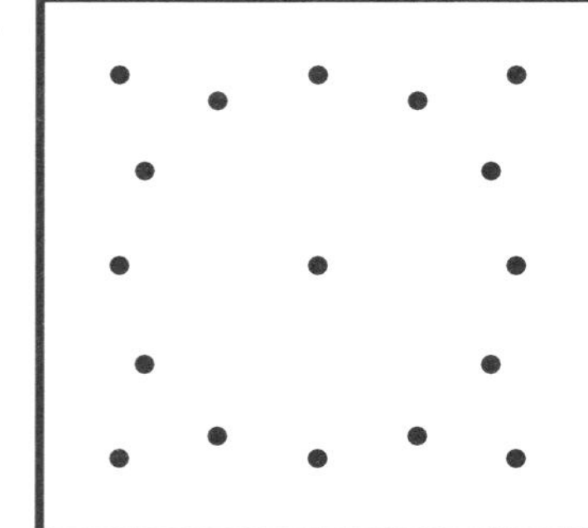

2.

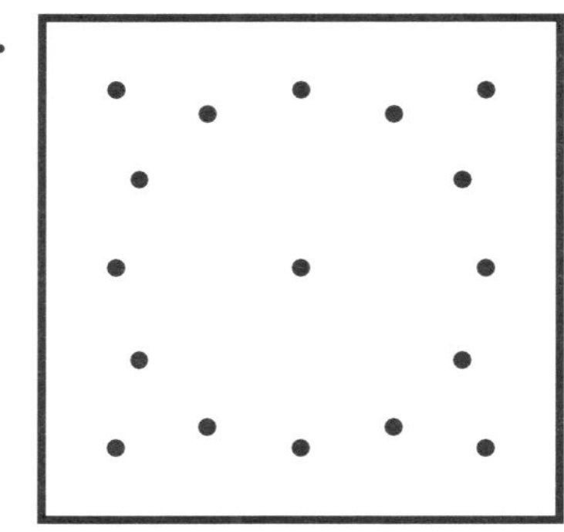

3.

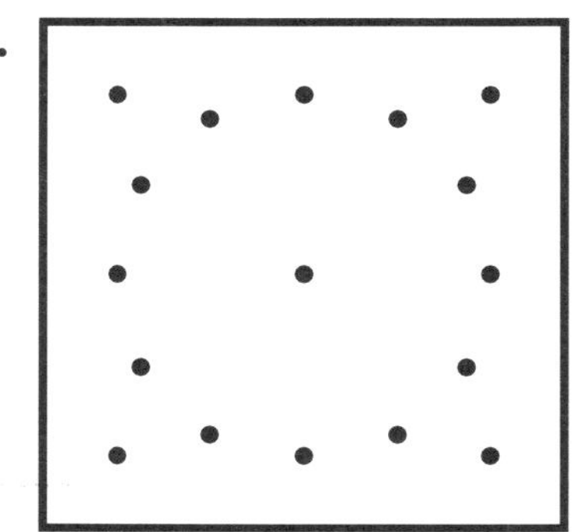

4.

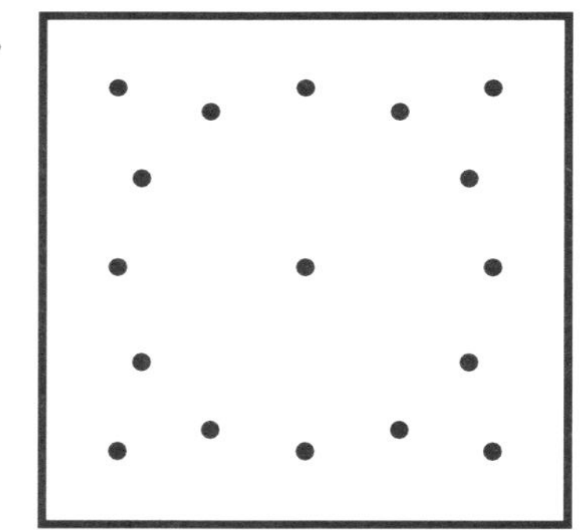

5.

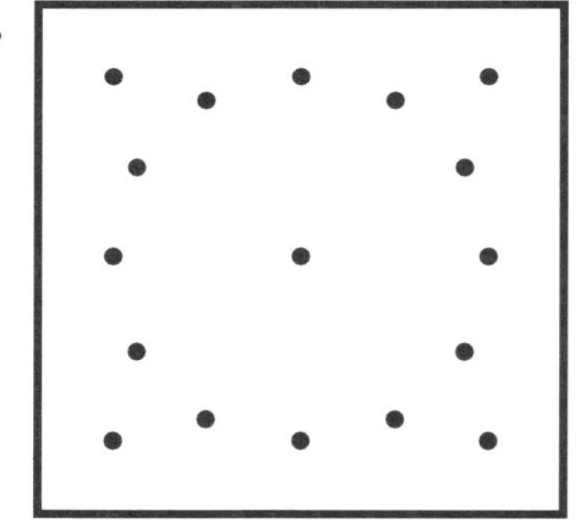

Name .... Date ....

Circumference and Diameter

The ***circumference*** is the distance around a circle.

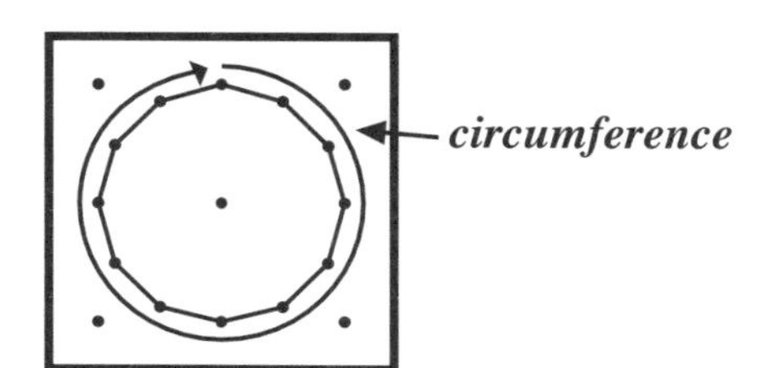

Do this activity with a partner to discover the relationship between the circumference and the diameter of a circle.

You will need: a circular geoboard, 2 feet of string, scissors, and a ruler.

Follow these steps:

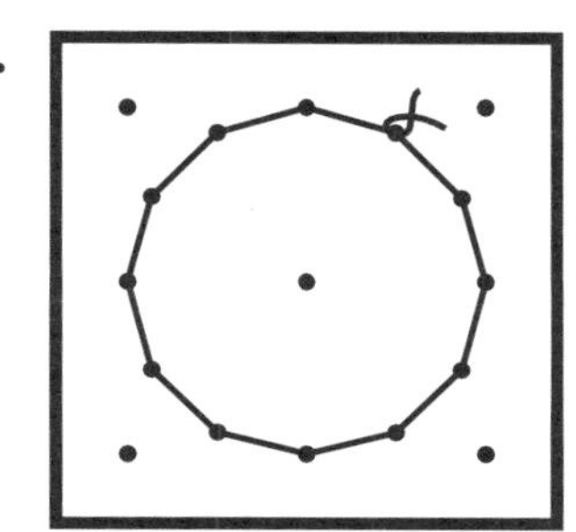

1. Tie a piece of string around the 12 pins that form the circle. This is the circumference. Cut off the ends of the string at the knot.

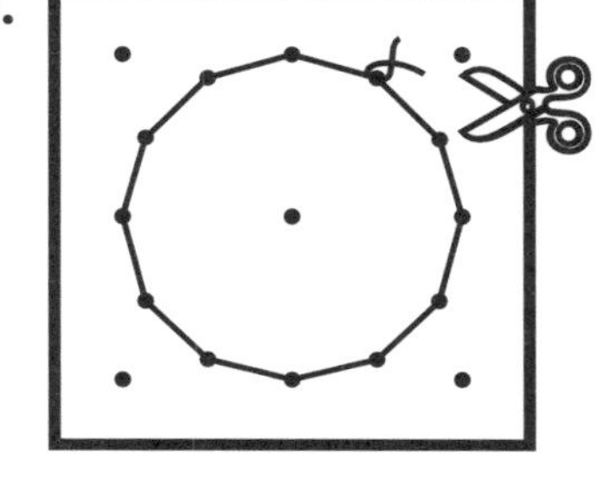

2. Cut the string from the pins to show one long piece of string.

3. Predict how many times do you think this piece of string could be placed along the diameter of your geoboard. ______

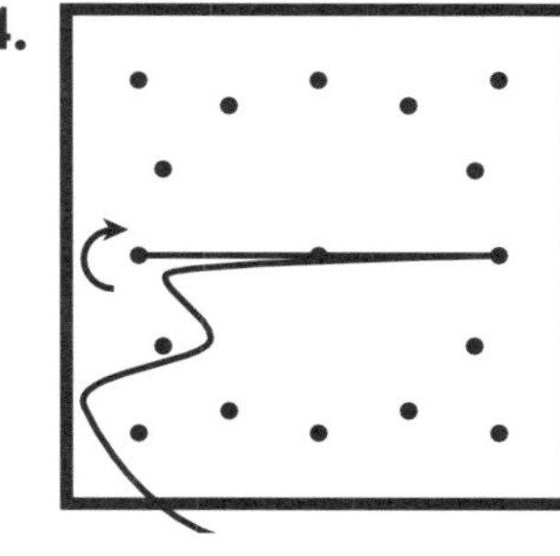

4. Place the string along the diameter on your geoboard as many times as you can. How many times did the piece of string go over the diameter? ______

How close was your prediction? ____________________________

5. Measure the piece of string with a ruler. How long is it? _______

Measure the length of the geoboard's diameter. How long is it? ______

Divide the length of the string by the length of the diameter. What do you get? ______

Is the answer close to the answer in Step 4 above? ______

6. Measure the circumference and diameter of at least three round objects.
Divide the circumference by the diameter for each.
What do you notice about the answer for each division? __________________________

Fractions and Angle Measure

Vocabulary

acute angle, central angle, central angle triangle, degree, obtuse angle, right angle, straight angle

Activity Teaching Notes

Central Angles and Fractions *(page 41)*
Before students model the unit fractions to show same-size central angles, have them complete this set of equivalent fractions:

$\frac{1}{2}$ = _____ /12 $\frac{1}{3}$ = _____ /12 $\frac{1}{4}$ = _____ /12 $\frac{1}{6}$ = _____ /12

This will help them model the unit fractions on their geoboards knowing how many $\frac{1}{12}$s will make a specific central angle and its arc. For example, $\frac{3}{12}$ will show the central angle for $\frac{1}{4}$.

Measures of Central Angles *(page 42)*
The activity on this page relates same-size fraction parts of a circle to the measure of central angles that can be modeled on a 12-pin geoboard. A central angle is measured in degrees. The smallest central angle on this geoboard is 30 degrees which is $\frac{1}{12}$ of the circle. Students can verify the measure of the central angles by using a protractor. Note: Angle measurement with protractors may result in slight variation due to the somewhat inexact nature of the geoboard.

Types of Central Angles *(page 43)*
Modeling acute, right, obtuse, and straight angles on a circular geoboard, students can literally see how large each angle is compared to modeling them on a 5 x 5-pin geoboard. Knowing the measure of the smallest acute angle that can be formed on the circular geoboard (30°) is a good benchmark for figuring out the type and size of angles for the rest of the angles on the page. Ask students how the number of pins can help them figure out how to model the angles.

Central Angle Triangles *(page 44)*
Knowing how to make a central angle with a certain number of degrees enables students to make a central angle triangle knowing the measure of one of the angles of the triangle. Also note that two sides of each central angle triangle are radii. Problems 6-11 are important in that they help students figure out the number of degrees for the remaining angles of a central angle triangle. Ask them what part of the circle forms two sides of the triangle, and why a central angle triangle cannot be scalene.

Central Angles and Fractions

1. Show all the diameters on your geoboard using six rubber bands. Record them at the right.

2. How many same-size central angles are there? ______

3. What fractional part of the circle is each of these central angles? ______

4. What would be the lowest-terms fraction for 6 of these central angles? ______ (Hint: $\frac{6}{12}$ = __/__)

for 3 of these central angles? ______

5. How can the pins on the geoboard help you know the fraction represented by a central angle? ______________________________

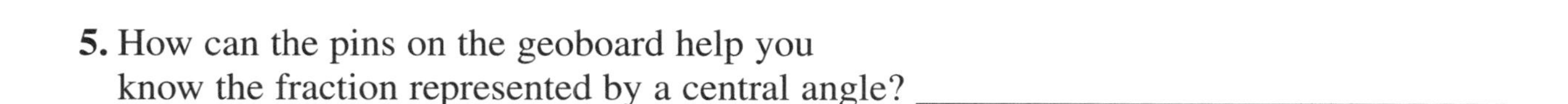

Model the central angle and its arc for each unit fraction on your geoboard. Record them below.

6. $\frac{1}{12}$

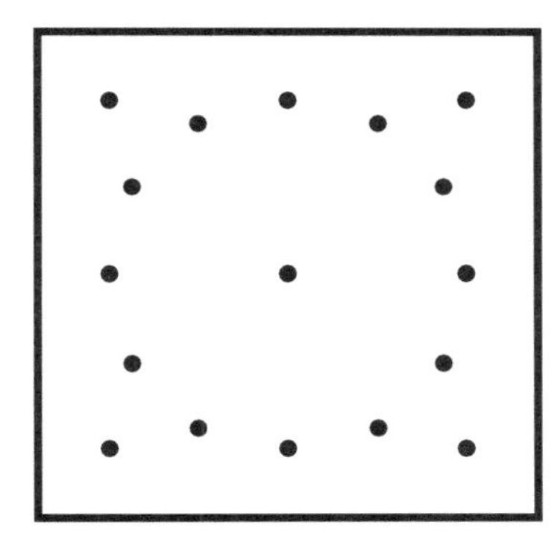

7. $\frac{1}{6}$

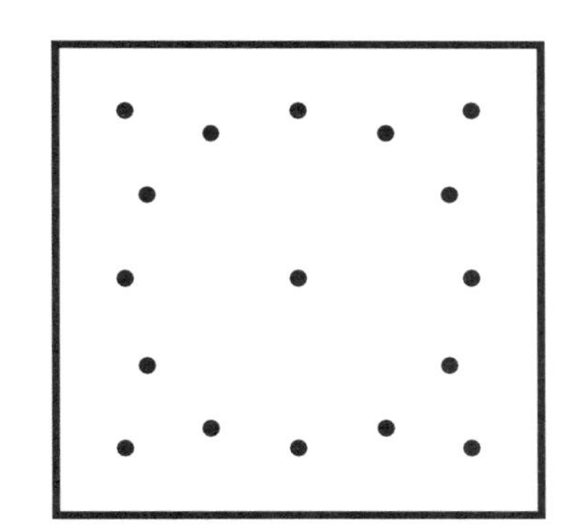

8. $\frac{1}{4}$

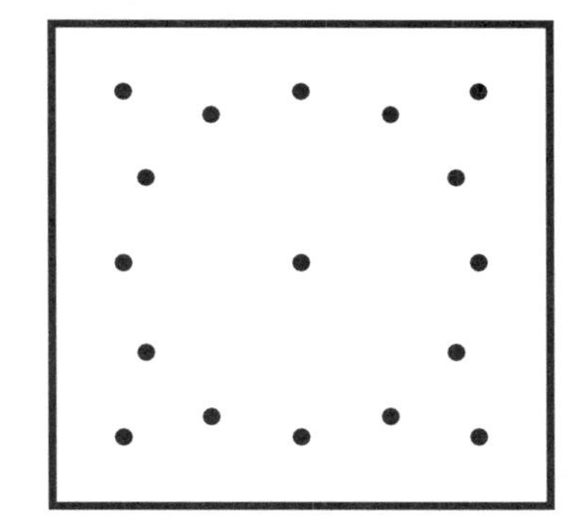

9. $\frac{1}{3}$

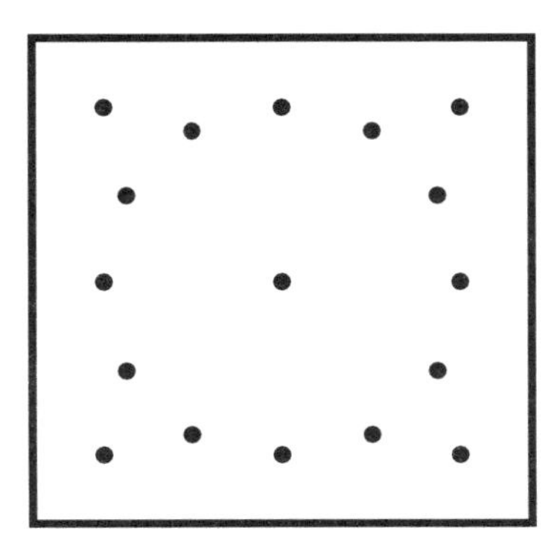

10. $\frac{1}{2}$

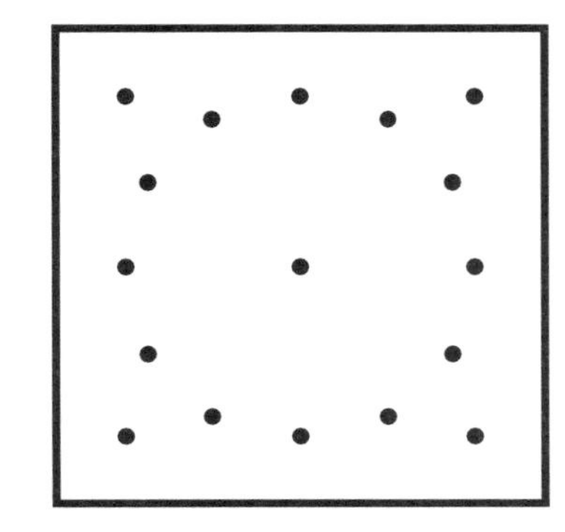

Name .... Date ....

Measures of Central Angles

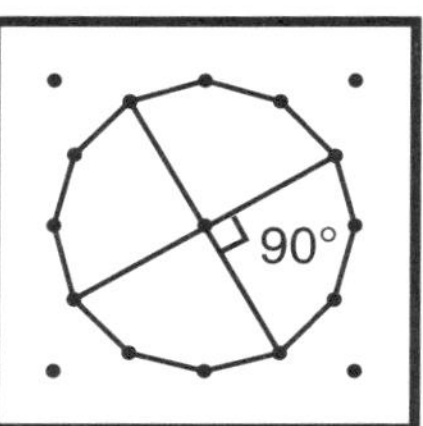

360° ÷ 4 = 90°

There are 360 ***degrees*** (°) in a circle. Fractional parts of a circle can tell you how many degrees are in a central angle. For example: If you have 4 same-size angles, divide 360° by 4 to find the measure of one of the central angles.

Model and record same-size central angles on your geoboard.
Then figure out the measure of the same-size central angle of each circle.

1. Model 2 same-size parts.

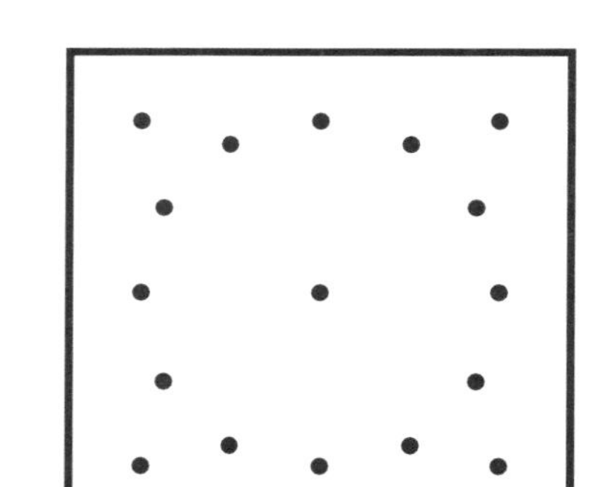

360° ÷ _____ = _____°
Each central angle is _____°

2. Model 6 same-size parts.

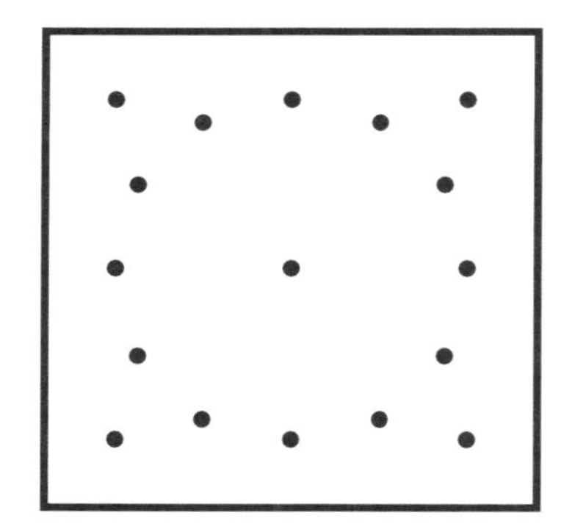

360° ÷ _____ = _____°
Each central angle is _____°

3. Model 3 same-size parts.

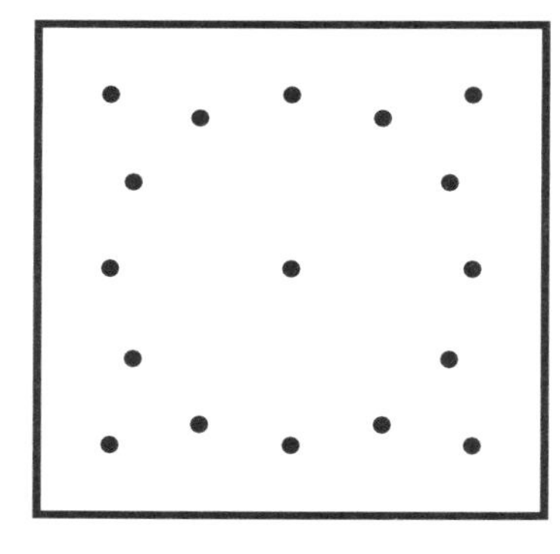

360° ÷ _____ = _____°
Each central angle is _____°

4. Model 4 same-size parts.

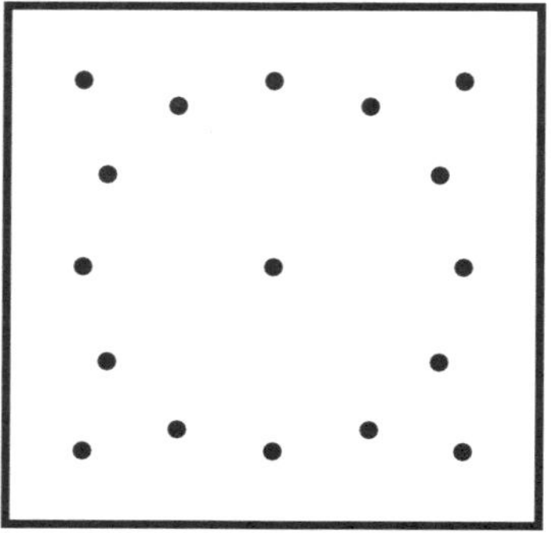

360° ÷ _____ = _____°
Each central angle is _____°

5. Model 12-same-size parts.

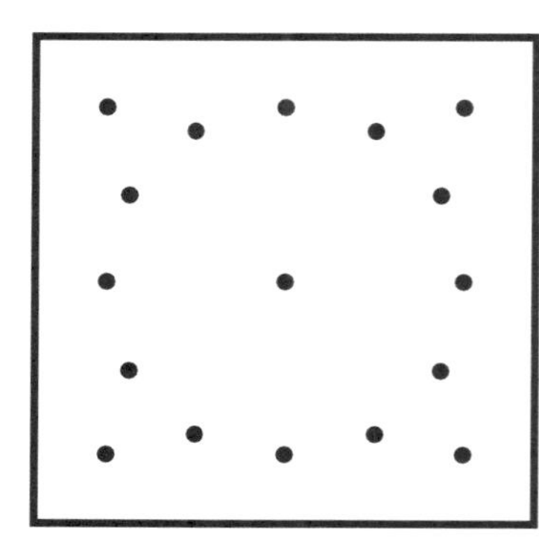

360° ÷ _____ = _____°
Each central angle is _____°

6. Model ⅛s.

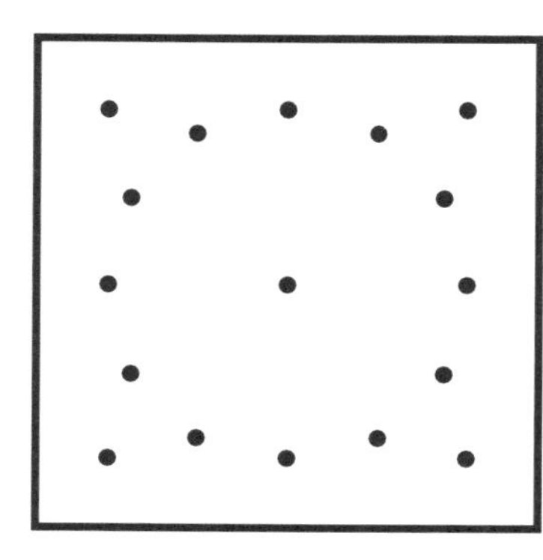

360° ÷ _____ = _____°
Each central angle is _____°

Name .... Date ....

Types of Central Angles

An ***acute angle*** has a measure between 0 and 90 degrees.

A ***right angle*** has a measure of 90 degrees.

An ***obtuse angle*** has a measure between 90 and 180 degrees.

A ***straight angle*** has a measure of 180 degrees.

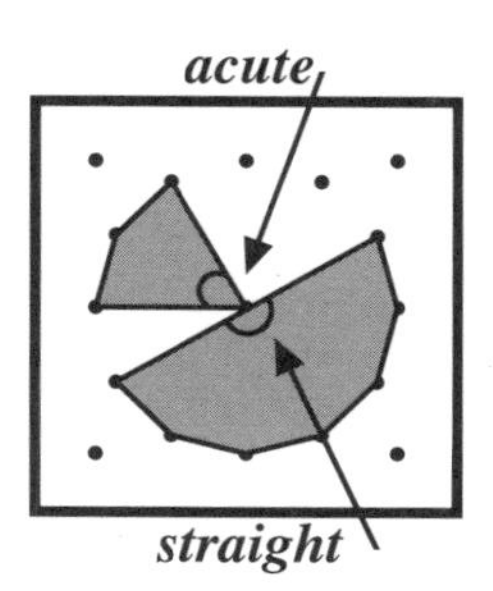

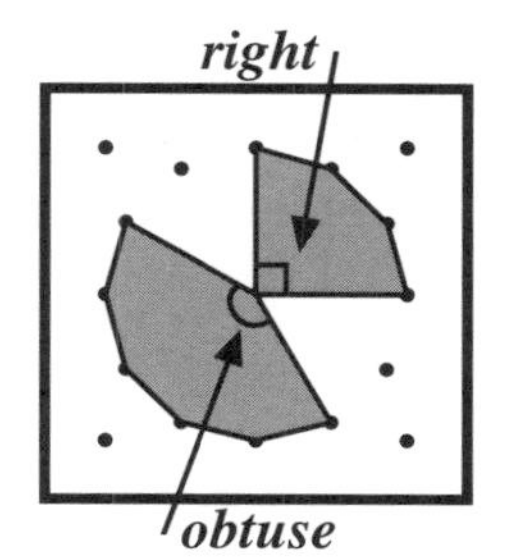

Model each central angle and its arc on your geoboard. Record them below.

1. Smallest acute angle

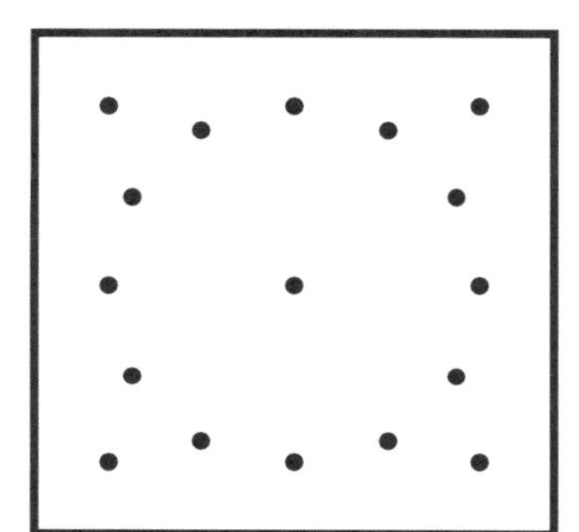

2. Right angle

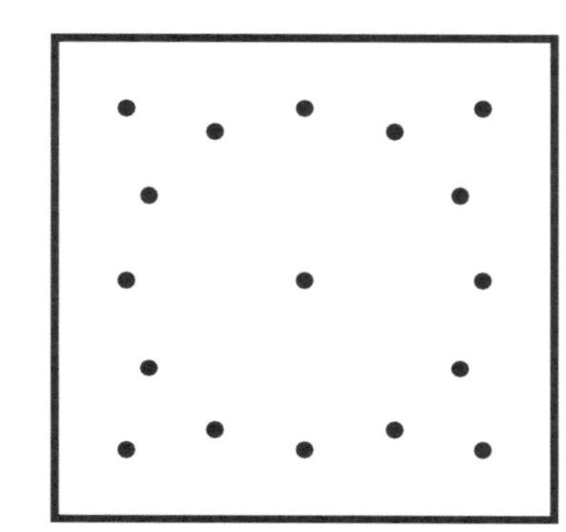

3. Straight angle

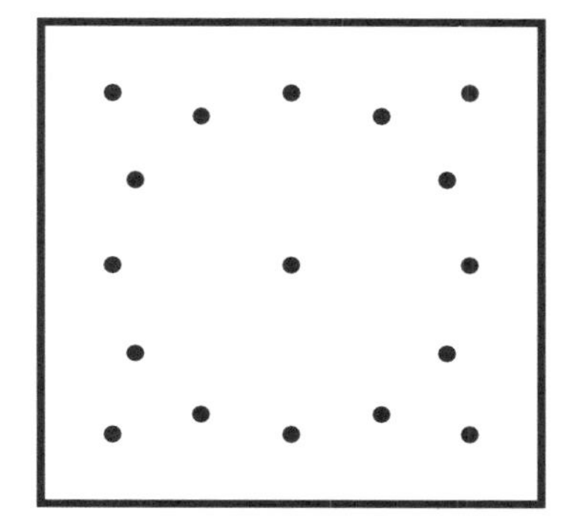

4. Smallest obtuse angle

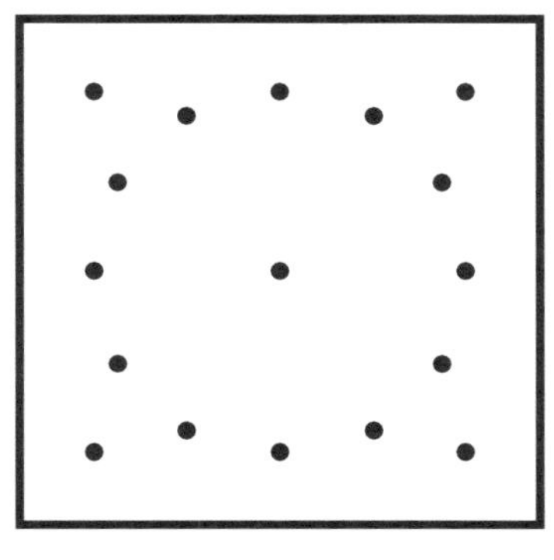

5. Largest acute angle

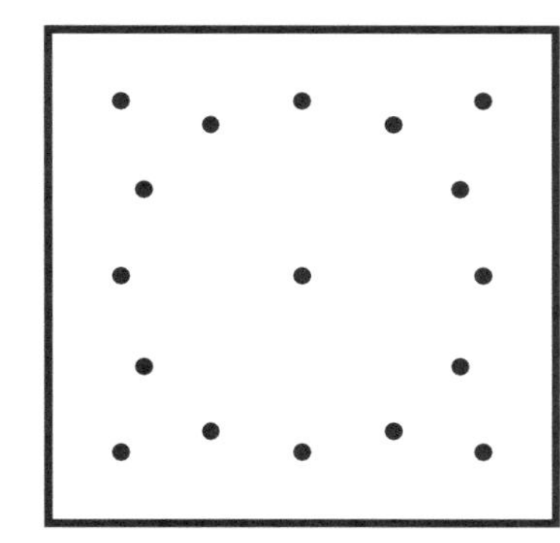

6. Largest obtuse angle

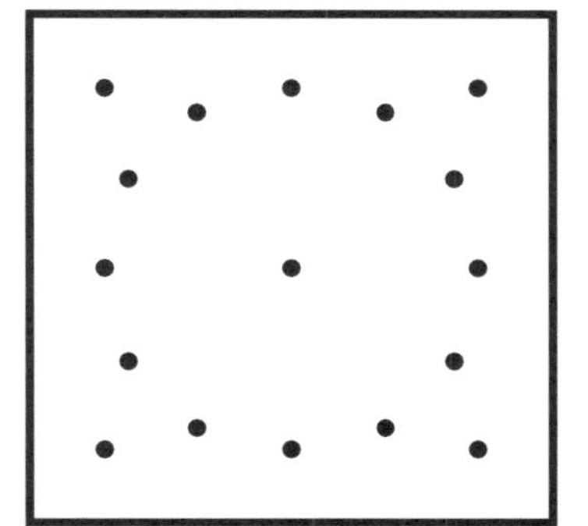

7. How many degrees are in each of the angles above?

#1: ______, #2: ______, #3: ______,

#4: ______, #5: ______, #6: ______

Name .... Date ....

Central Angle Triangles

A ***central angle triangle*** is formed with a central angle of a circle and a chord connecting the two endpoints of the central angle on the circle.

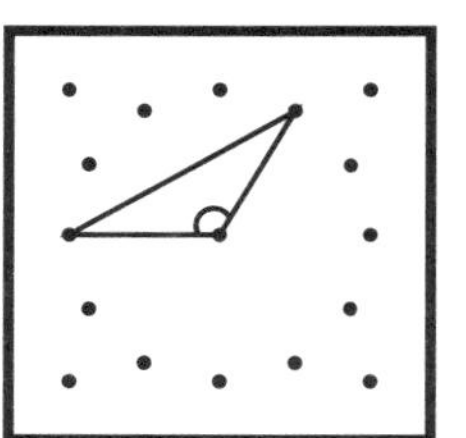

120° central angle triangle

Model the following central angle triangles on your geoboard. Record them below.

1. 30° central angle triangle

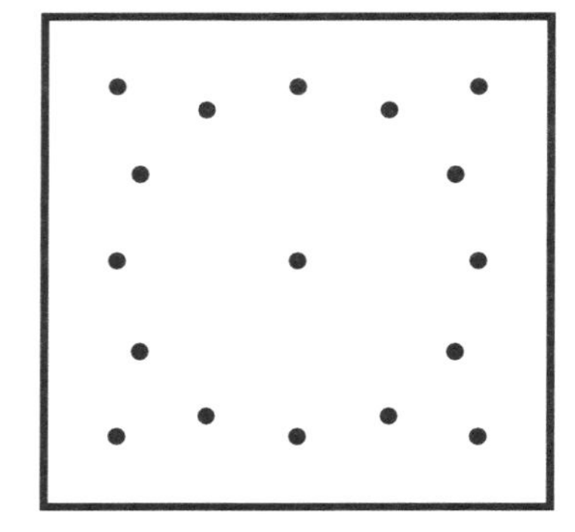

2. 90° central angle triangle

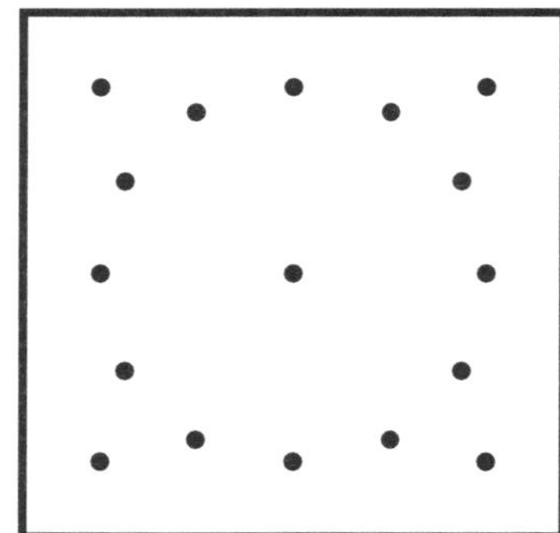

3. 150° central angle triangle

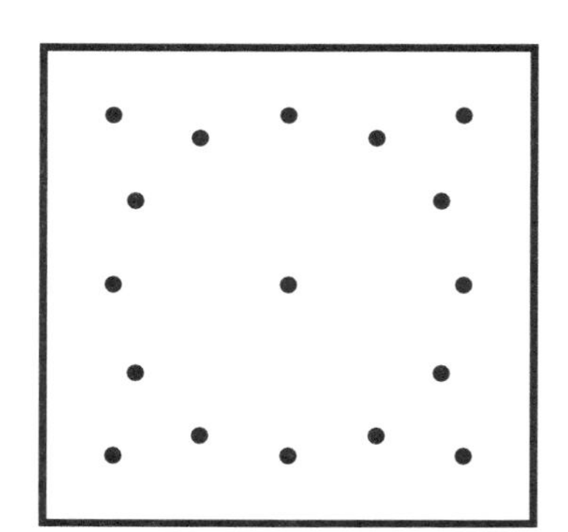

4. 60° central angle triangle

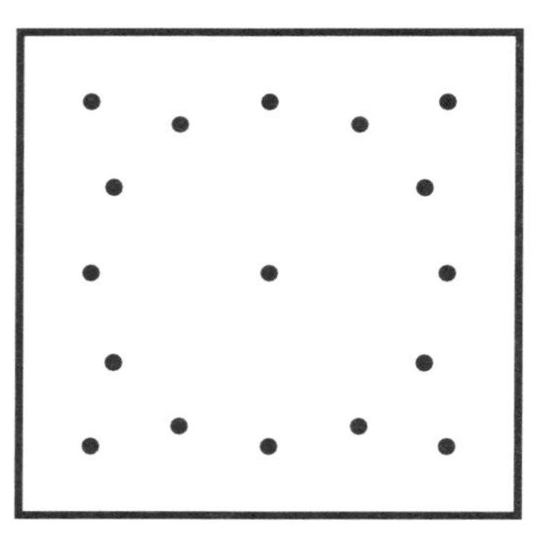

5. 120° central angle triangle

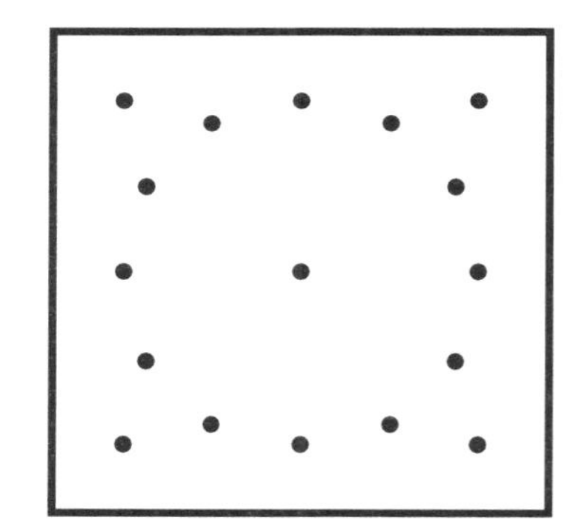

Write the numbers of the Problems (1-5) above in the blanks to answer the questions.

6. Which triangles are acute triangles? ______

7. Which triangles are right triangles? ______

8. Which triangles are obtuse triangles? ______

9. Which triangles are equilateral? ______

10. Which triangles are isosceles? ______

11. Which triangles are scalene? ______

Inscribed Figures

Vocabulary

inscribed angle, inscribed polygon, inscribed quadrilateral, inscribed triangle

Activity Teaching Notes

Inscribed Angles *(page 46)*
An inscribed angle is an angle whose vertex in on the circle and its sides are chords, unlike a central angle whose vertex is the center of the circle and each side is a radius. Ask students to explain the difference between an inscribed angle and a central angle. Ask students if some of their inscribed angles have a diameter as one of the sides. Give students copies of the circular recording sheets, and challenge them to find and record more inscribed angles on their geoboards.

Measures of Inscribed Angles *(page 47)*
Since the measure of an inscribed angle is one-half the measure of the central angle cut off by the same arc, students should model both angles. Use two rubber bands, one to model the inscribed angle with its arc and one to model the central angle with the same arc. After students complete the page they can verify the inscribed and central angle measures with a protractor.

Inscribed Triangles *(page 48)*
This inscribed triangle activity also contains a review of different types of triangles in terms of angle measures and lengths of their sides. Refer back to page 15. Provide students with copies of circular recording sheets. Have them find all the possible inscribed acute, obtuse, and right triangles on their geoboards and record them. They should notice that in an inscribed right triangle, the diameter is always one of its sides.

Inscribed Polygons *(page 49)*
Several solutions are possible for these problems. Have students work in small cooperative groups to find as many of them as they can. Give them extra copies of circular recording sheets to record the solutions. Also urge them to model polygons with 3 and 4 sides. Ask students if they formed any regular polygons and if there is a pattern in the pins to model it. For an equilateral triangle, the vertices are at every fourth pin. For a square, the vertices are at every third pin. For a regular hexagon, the vertices are at every second pin.

Name .... Date ....

Inscribed Angles

An ***inscribed angle*** has its vertex on the circle. Each side is a chord.

Model 8 different inscribed angles with their arcs on your geoboard. Record them below.

1.

2.

3.

4.

5.

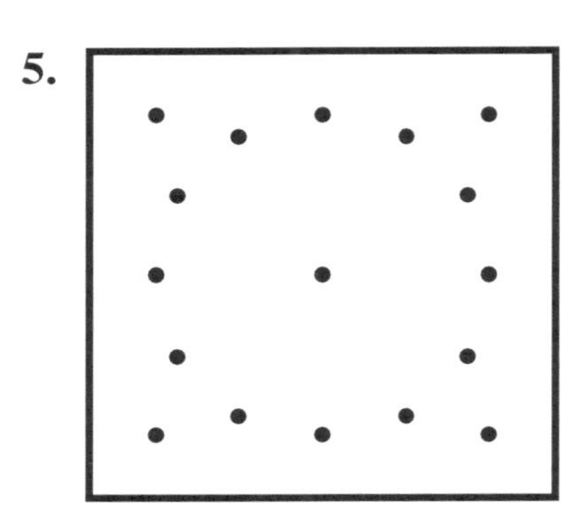

6.

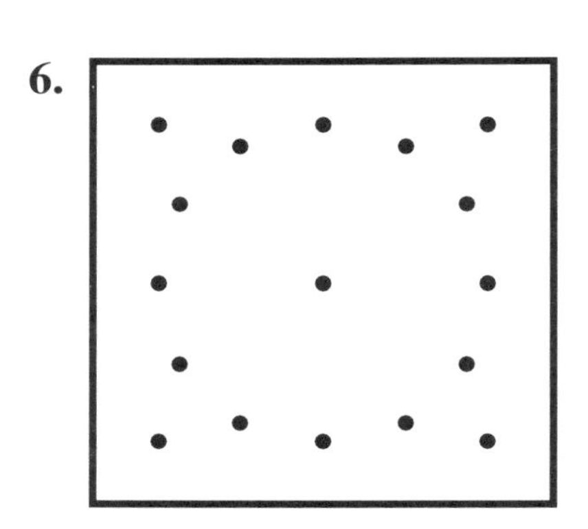

7.

8.

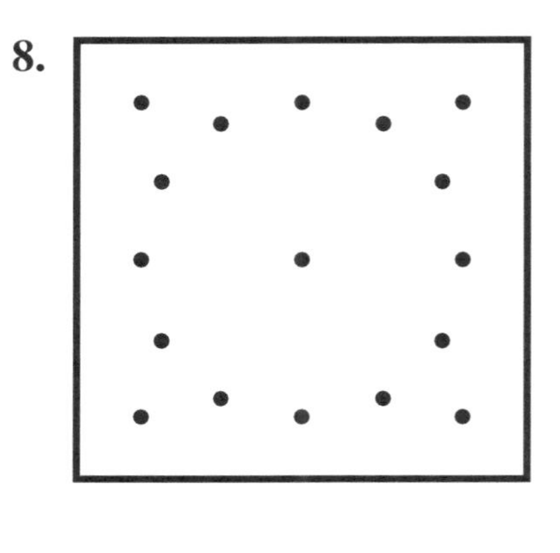

Name .... Date ....

Measures of Inscribed Angles

The measure of an inscribed angle is one-half the measure of the central angle that cuts off the same arc.

central angle

Inscribed angle is one-half the measure of the central angle.

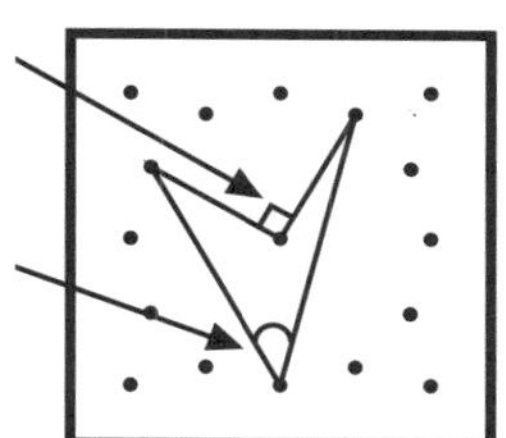

Step 1: Model the central angle that cuts the same arc. Central angle is 90°
Step 2: Divide the measure of the central angle by 2. Inscribed angle is 90° ÷ 2 = 45°

Model each inscribed angle with arc on your geoboard.
Then model the central angle cut by the same arc.
Figure our the measure of the inscribed angle.

1. Central angle is ______.

Inscribed angle is ______.

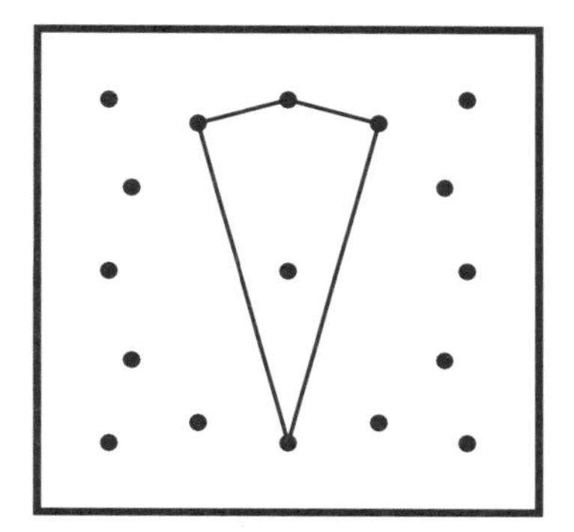

2. Central angle is ______.

Inscribed angle is ______.

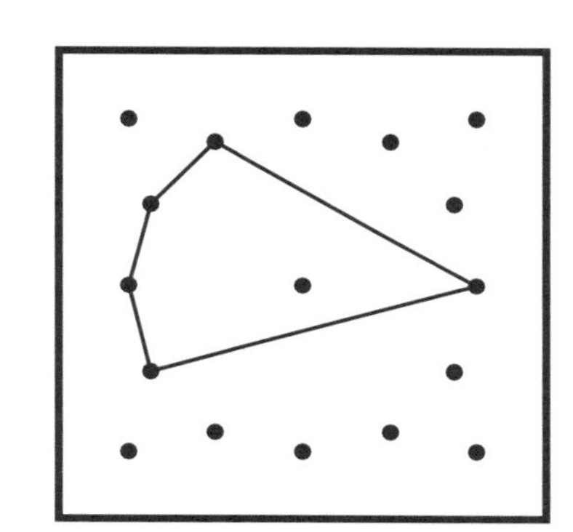

3. Central angle is ______.

Inscribed angle is ______.

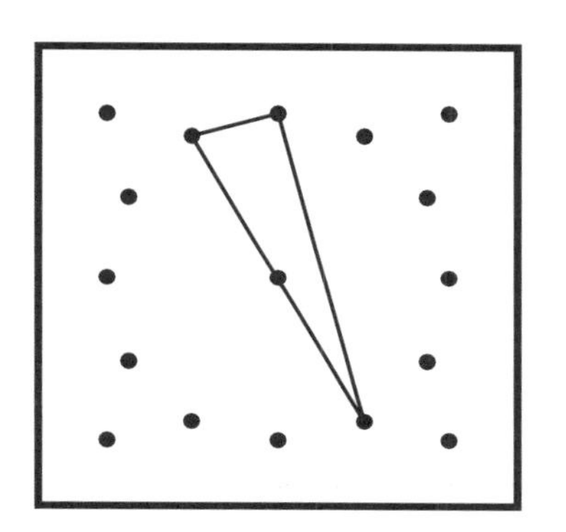

4. Central angle is ______.

Inscribed angle is ______.

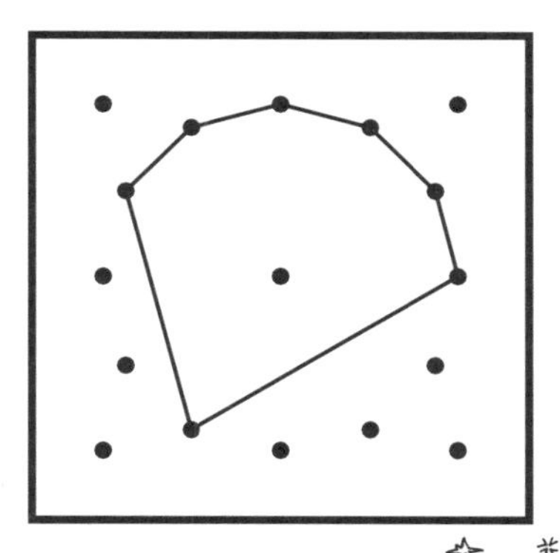

5. Central angle is ______.

Inscribed angle is ______.

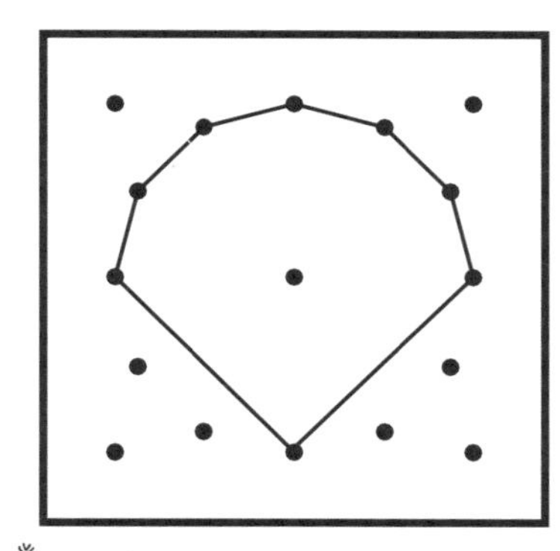

6. Central angle is ______.

Inscribed angle is ______.

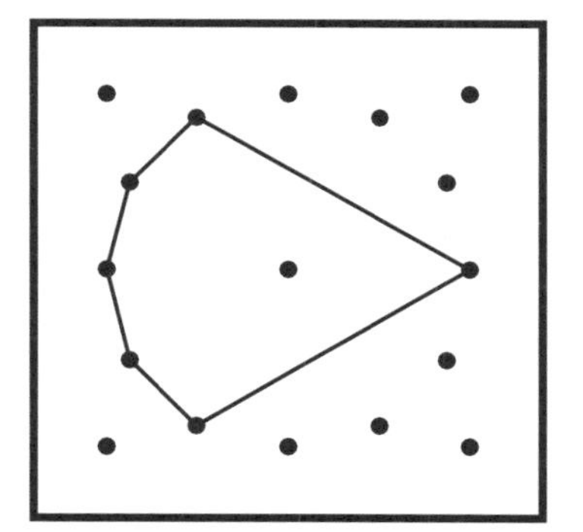

Inscribed Triangles

An ***inscribed triangle*** has all three of its vertices on the circle. Each side is a chord.

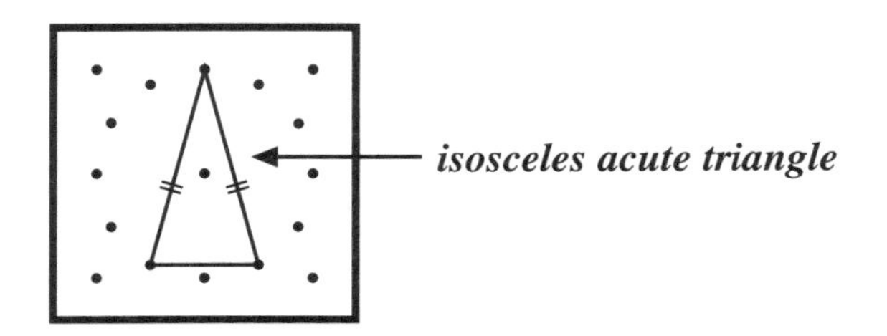

Model the following types of inscribed triangles on your geoboard. Record them below.

1. Isosceles acute triangle

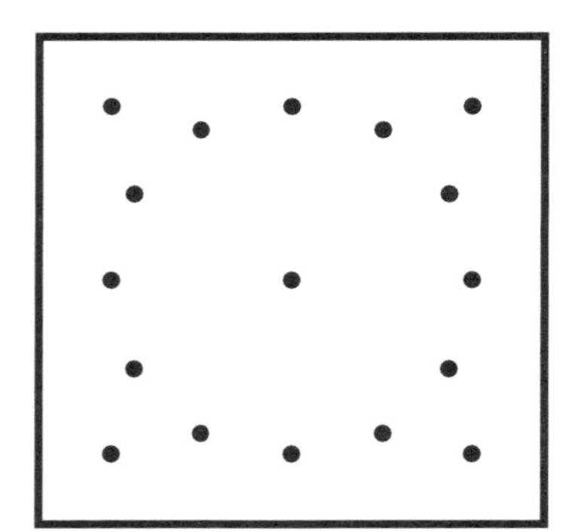

2. Isosceles right triangle

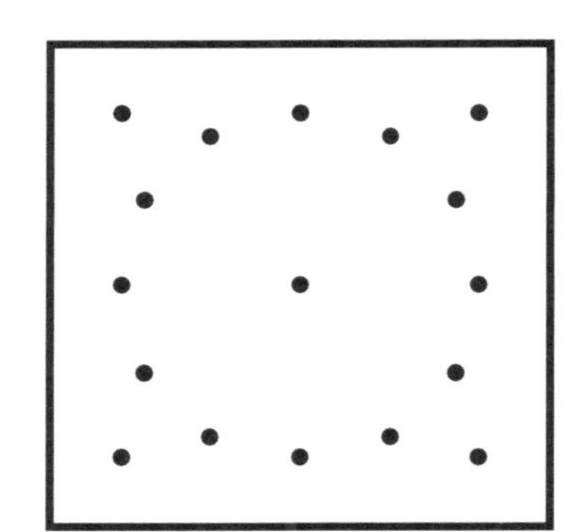

3. Equilateral triangle

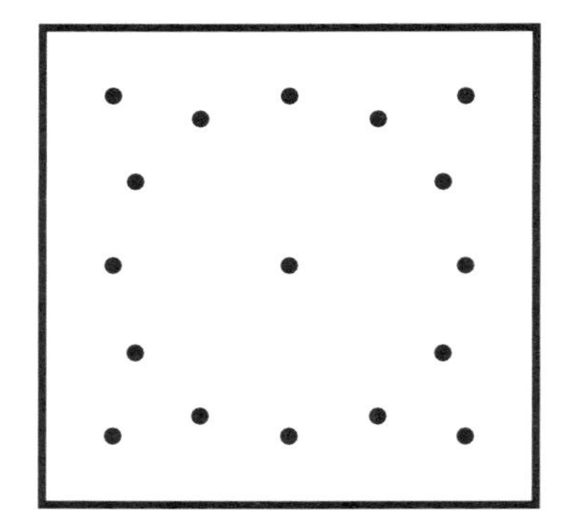

4. Scalene right triangle

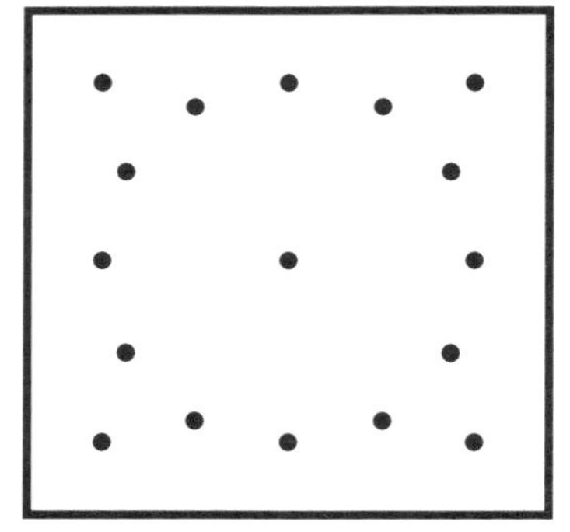

5. Isosceles obtuse triangle

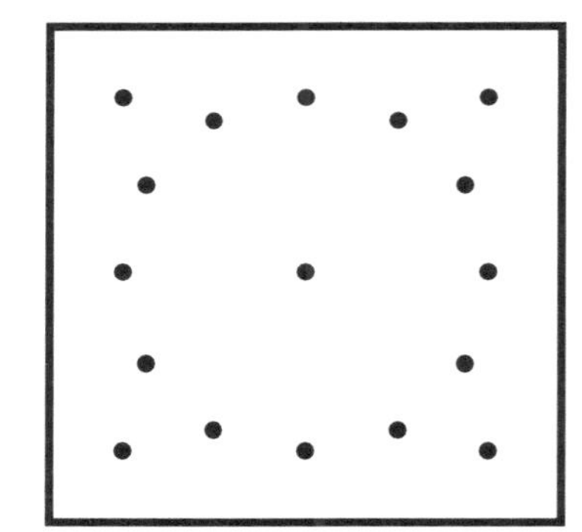

6. Scalene triangle

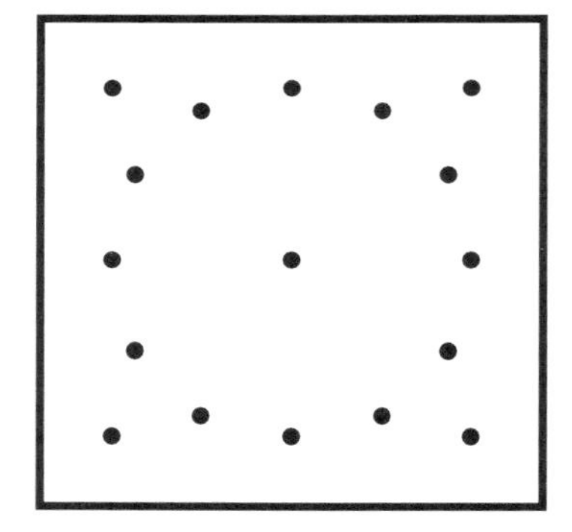

Name .... Date ....

Inscribed Polygons

An ***inscribed polygon*** has at least three sides with all vertices on the circle. Each side is a chord.

In an ***inscribed regular polygon,*** all sides are congruent.

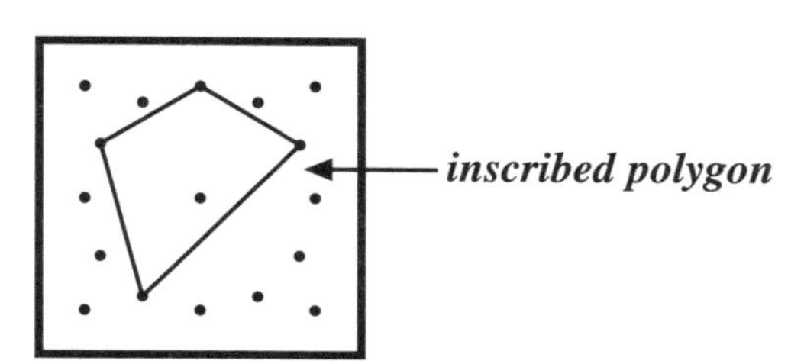

Model inscribed polygons on your geoboard with the indicated number of sides. Record them below. Circle any that are regular polygons.

1. 5 sides

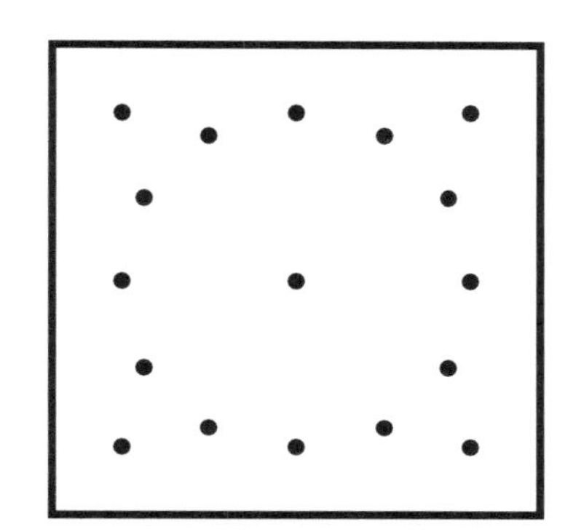

2. 6 sides

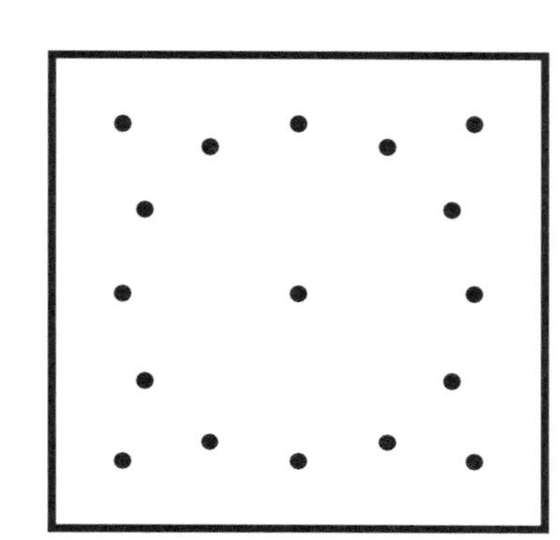

3. 7 sides

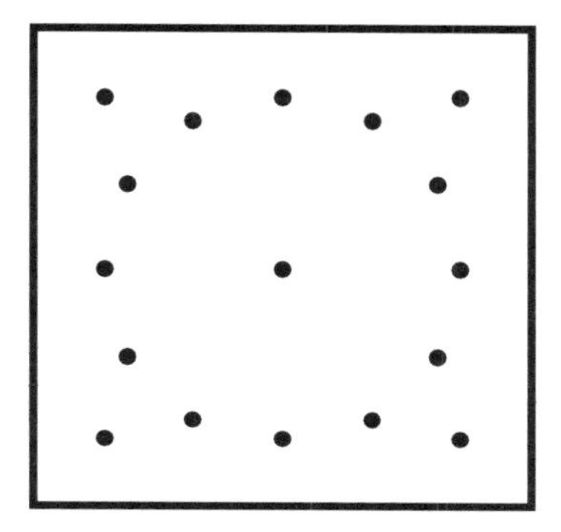

4. 8 sides

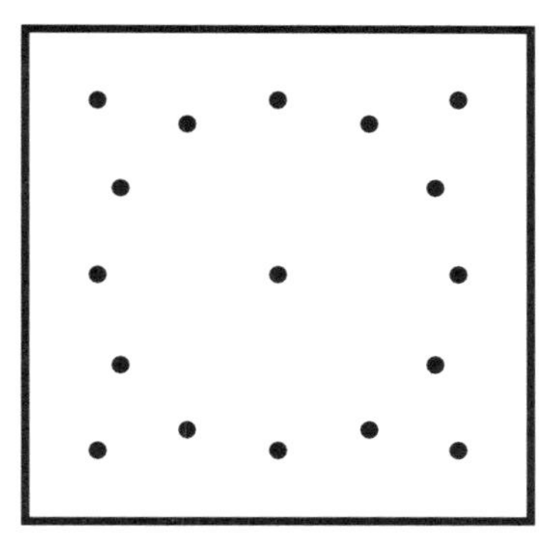

5. 9 sides

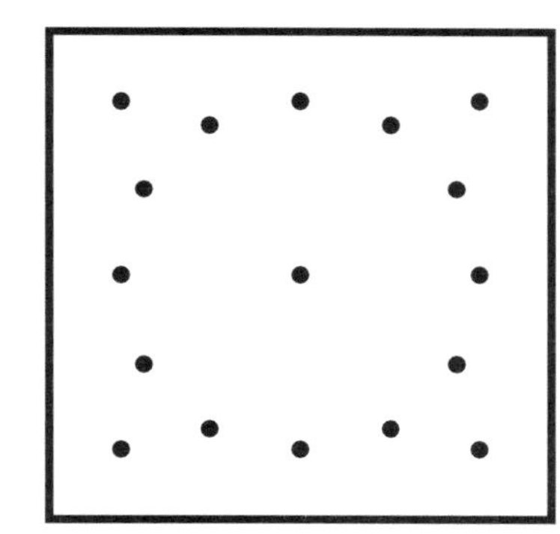

6. 10 sides

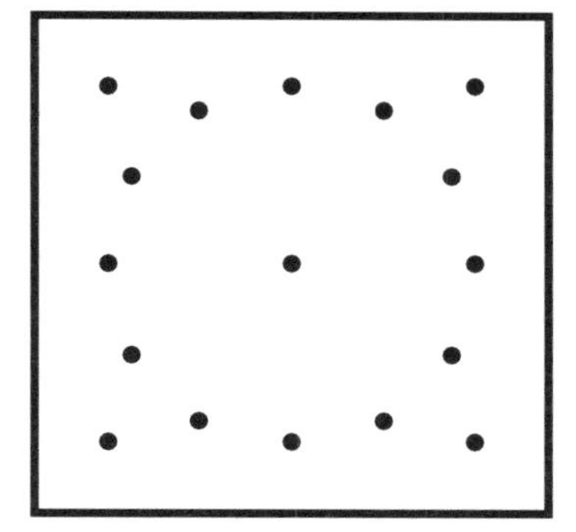

7. Can you name the type of polygons for the following problems?

#1: ____________________ #2: ____________________

#4: ____________________

Add Fractions

Vocabulary

central angle, fraction, lowest-terms fraction

Teaching Notes for Geoboard Activities

Model and Add Fractions ***(page 51)***
Adding fractions using a circular geoboard is like adding fractions with a set of fraction circle models consisting of $\frac{1}{2}$s, $\frac{1}{3}$s, $\frac{1}{4}$s, $\frac{1}{6}$s, and $\frac{1}{12}$s. Using a circular geoboard gives students a similar visual image but presents a different approach in that it requires them to model the addends. Have students model each fraction addend with a separate rubber band. For the problem $\frac{1}{4} + \frac{1}{2}$, show a central angle with its arc for $\frac{1}{4}$, and then show the central angle for $\frac{1}{2}$ with its arc adjacent to the central angle for $\frac{1}{4}$. The resulting figure should look like $\frac{3}{4}$ of the circular region, enclosed with rubber bands. Students may find equivalent fractions for each addend and add, or simply observe the model to see that the sum of $\frac{1}{4} + \frac{1}{2} = \frac{3}{4}$. Ask students how they arrived at their answers. Discuss Problems 7 and 8. Students should realize that between any two adjacent pins on the circle, the arc cut-off is opposite the central angle for $\frac{1}{12}$. They can count by $\frac{1}{12}$ths for each pin to model each addend and find each sum.

Model and Add More Fractions ***(page 52)***
This activity gives students more practice adding fractions using the circular geoboard. Discuss Problems 7-8. For Problem 7, the sum of $\frac{1}{2} + \frac{1}{2}$ is 1. Problem 8 involves adding three fractions. Ask the class to share how they found the sum.

Write and Solve Fraction Problems ***(page 53)***
Students are to write fractions for the central angles modeled on the geoboards. Then they are to find the sum. Discuss Problem 7. Some students may give lowest-terms fractions for each addend while others may not. Encourage them to write each problem using lowest-term fractions.

Name .... Date ....

Model and Add Fractions

Use two rubber bands, one for each central angle fraction, to model each addition problem on your geoboard. Record the model and the sum in lowest terms for each problem.

Example: Model, record: $\frac{1}{4} + \frac{1}{12} =$ ____

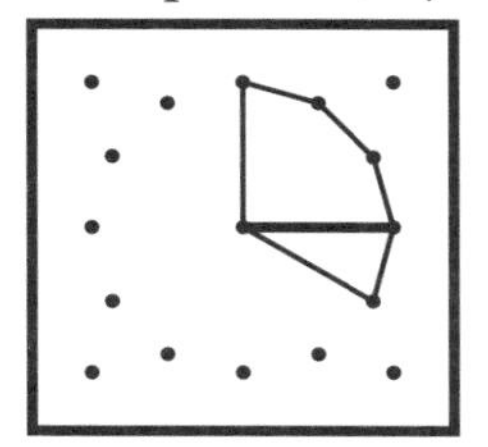

Think and do:
$\frac{3}{12} + \frac{1}{12} = \frac{4}{12} = \frac{1}{3}$

Write the sum:
$\frac{1}{4} + \frac{1}{12} = \frac{1}{3}$

1. $\frac{1}{2} + \frac{1}{4} =$ ______

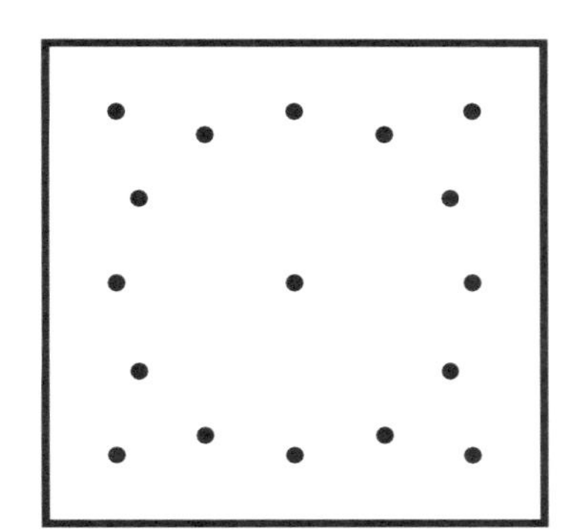

2. $\frac{1}{3} + \frac{1}{6} =$ ______

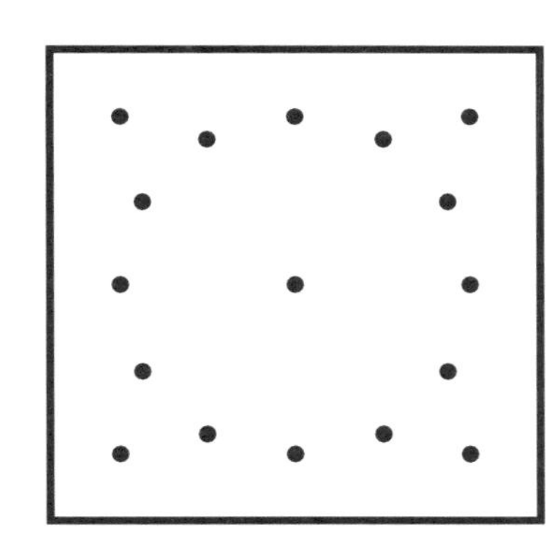

3. $\frac{1}{6} + \frac{1}{2} =$ ______

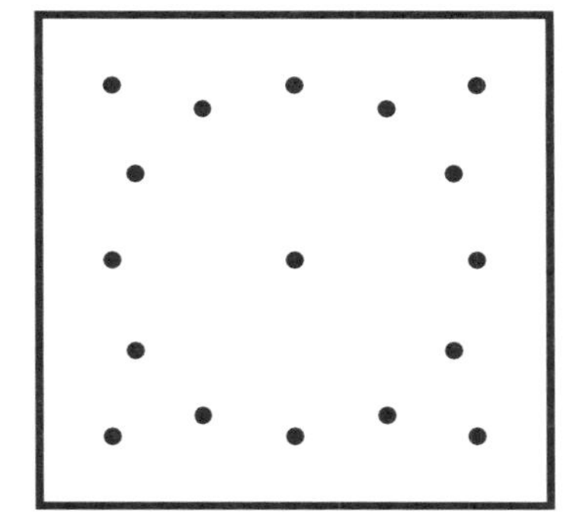

4. $\frac{5}{12} + \frac{1}{6} =$ ______

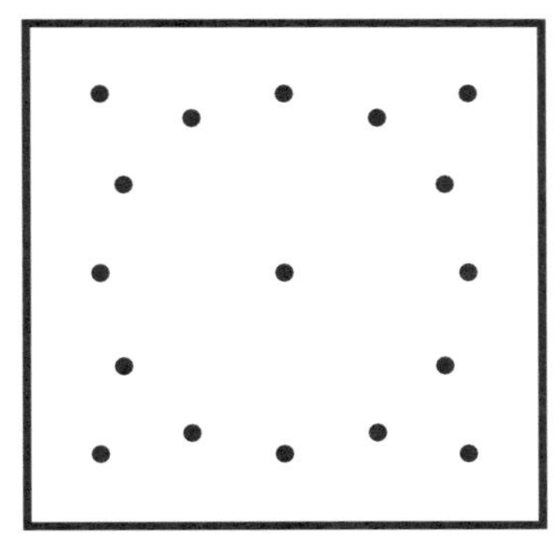

5. $\frac{1}{3} + \frac{1}{4} =$ ______

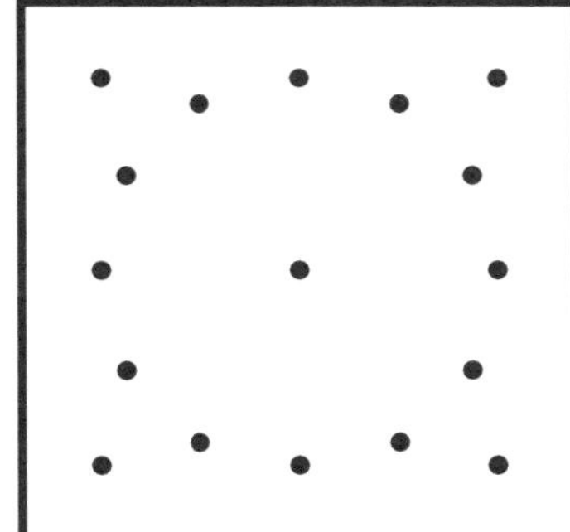

6. $\frac{7}{12} + \frac{1}{6} =$ ______

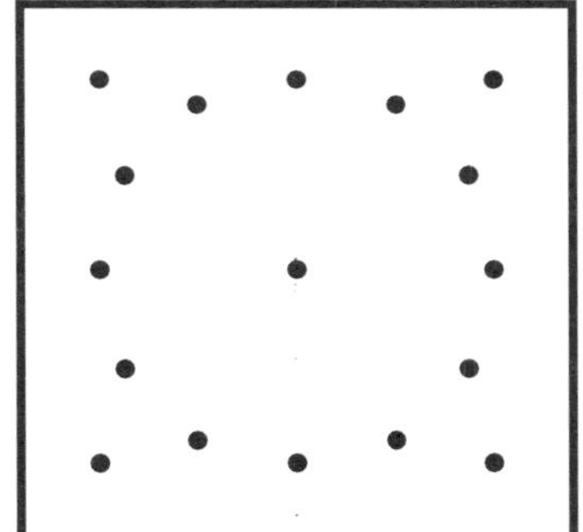

7. How did the pins on the geoboard help you model each fraction of the problem?

____

8. How did the pins on the geoboard help you find the sum of each problem?

____

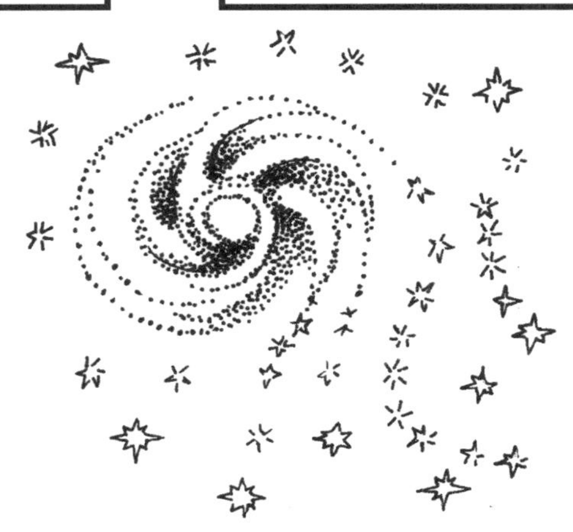

Name .... Date ....

Model and Add More Fractions

Use rubber bands, one for each central angle fraction, to model each addition problem. Record the model and the sum in lowest terms for each problem.

1. $\frac{1}{12} + \frac{1}{6} =$ _______

2. $\frac{1}{3} + \frac{1}{6} =$ _______

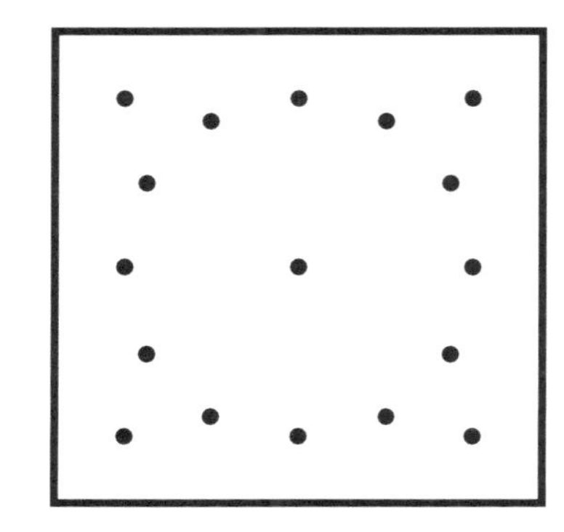

3. $\frac{1}{6} + \frac{1}{4} =$ _______

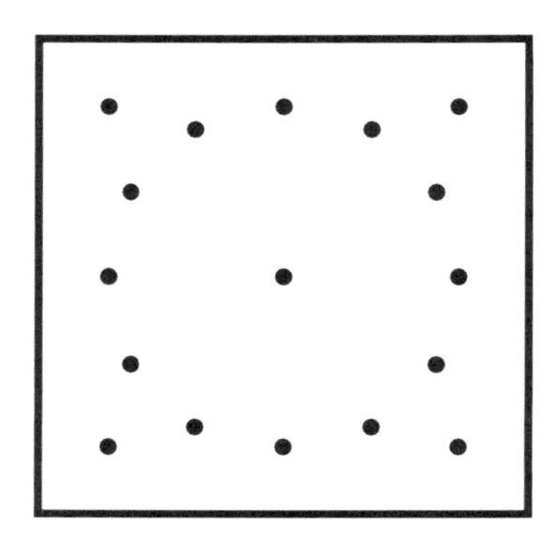

4. $\frac{1}{3} + \frac{1}{2} =$ _______

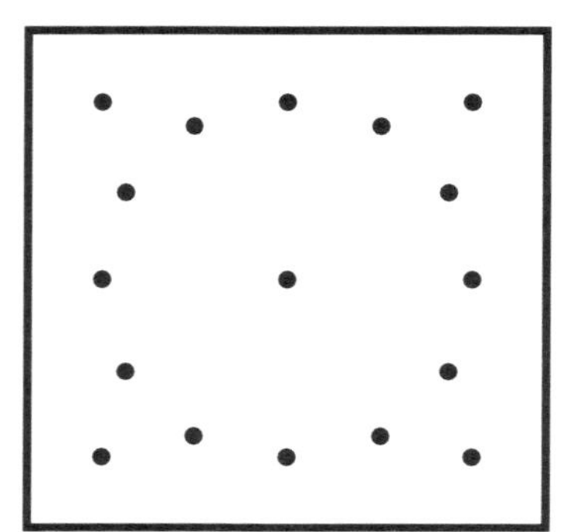

5. $\frac{1}{2} + \frac{1}{12} =$ _______

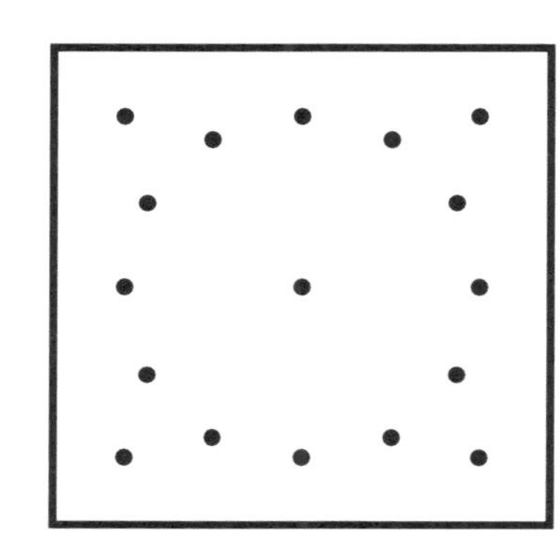

6. $\frac{1}{2} + \frac{1}{3} =$ _______

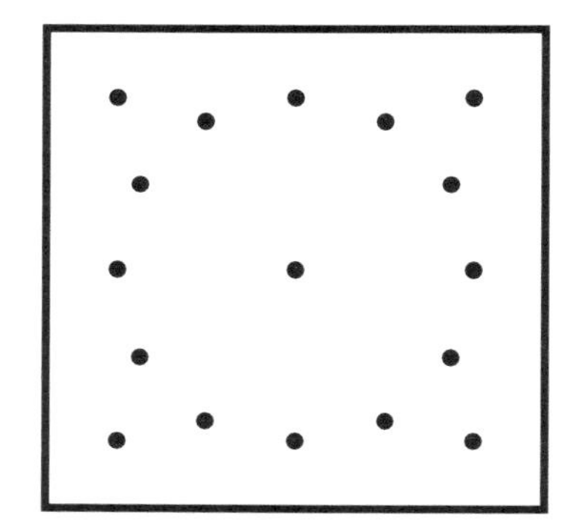

7. $\frac{1}{2} + \frac{1}{2} =$ _______

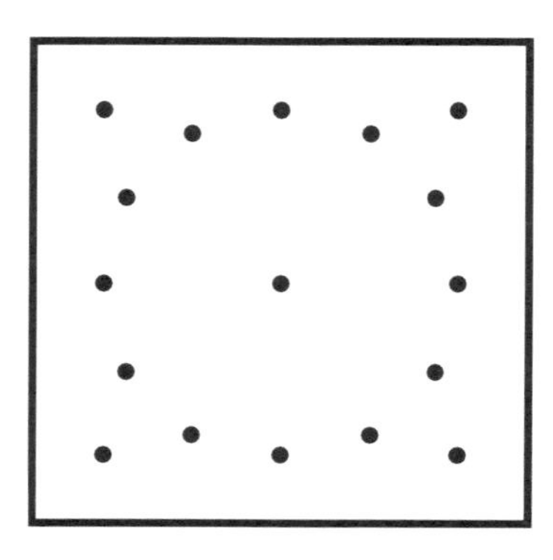

8. $\frac{1}{3} + \frac{1}{6} + \frac{1}{12} =$ _______

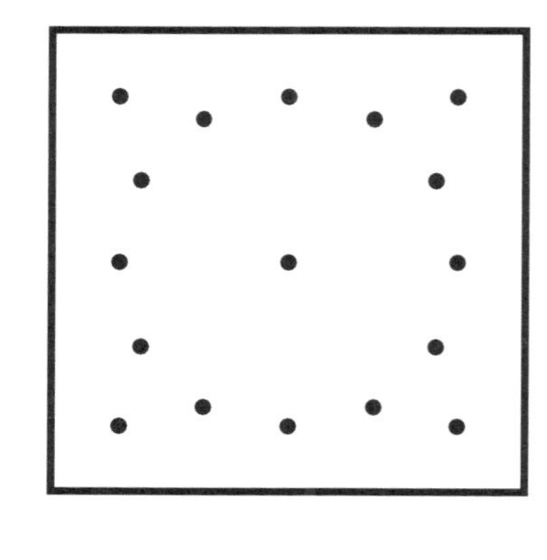

Name .... Date ....

Write and Solve Fraction Problems

Use two rubber bands. Model the addends showing central angle fractions. Record the addends and the sum below for each problem.

See and model:

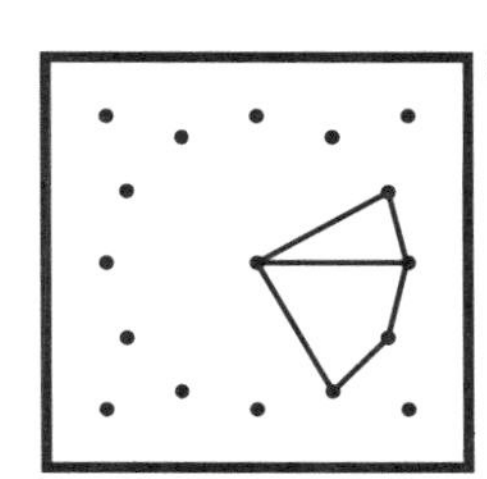

Think and do:

$\frac{1}{12} + \frac{1}{6} =$

$\frac{1}{12} + \frac{2}{12} = \frac{3}{12} = \frac{1}{4}$

Write:

$\frac{1}{12} + \frac{1}{6} = \frac{1}{4}$

1. ______ + ______ = ______

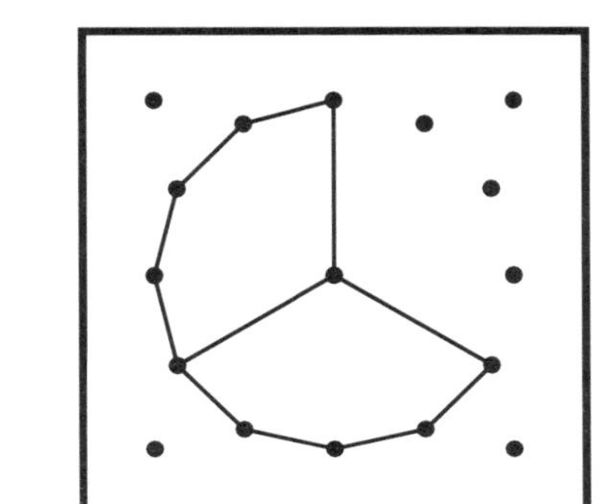

2. ______ + ______ = ______

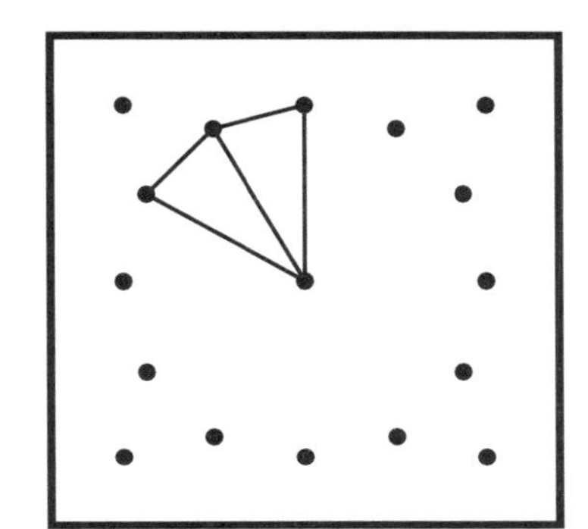

3. ______ + ______ = ______

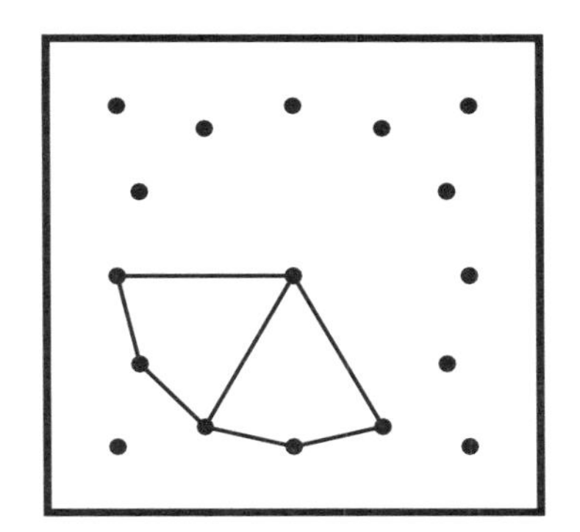

4. ______ + ______ = ______

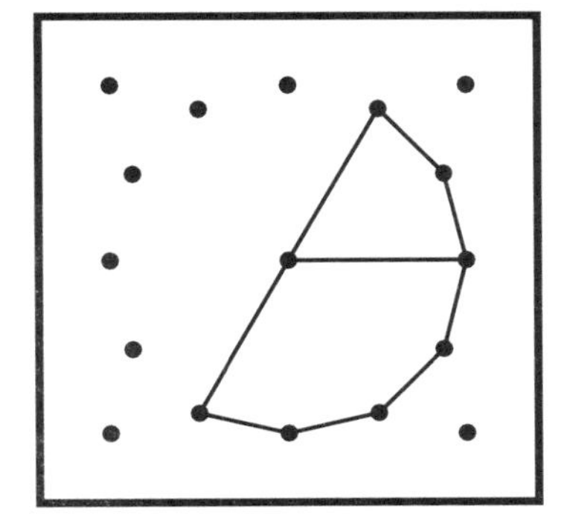

5. ______ + ______ = ______

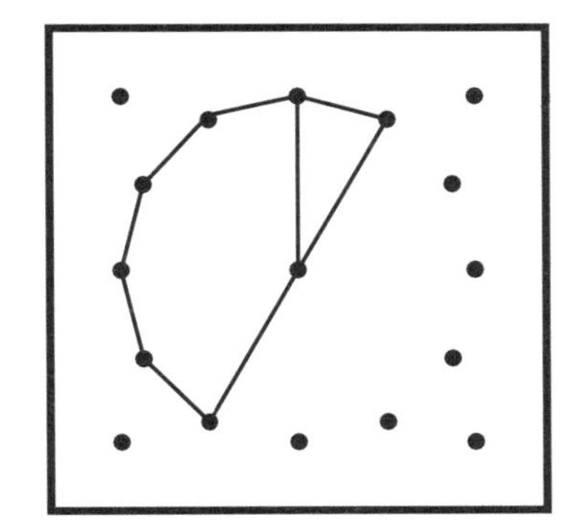

6. ______ + ______ = ______

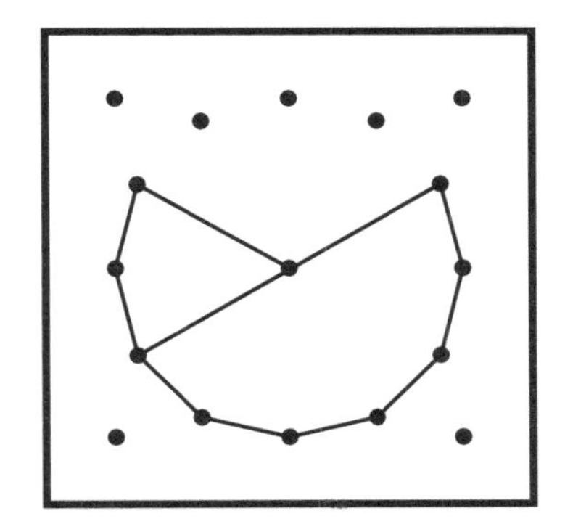

7. How did you know which fractions to write for the central angles in each problem?

____

Selected Solutions

Lines and Angles

Page 8:

2. 1, 2, 3, 4 **3.** 25, 50, 75, 100 **4–6.** Answers vary.

Page 9:

1. Some acute angles are shown.

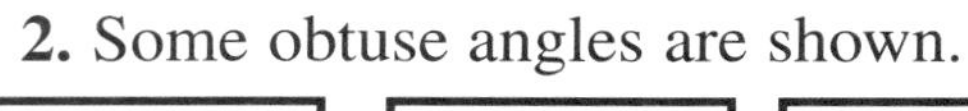

2. Some obtuse angles are shown.

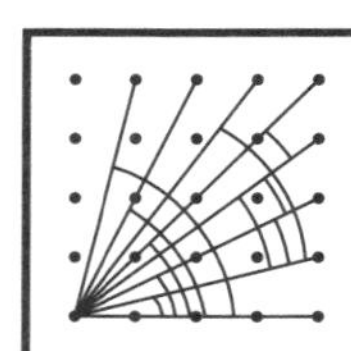
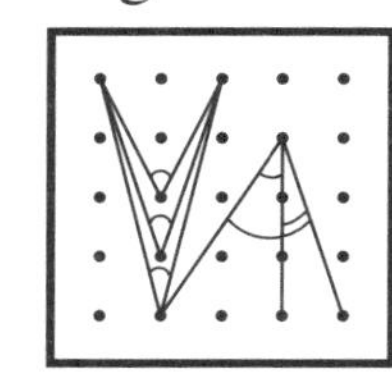
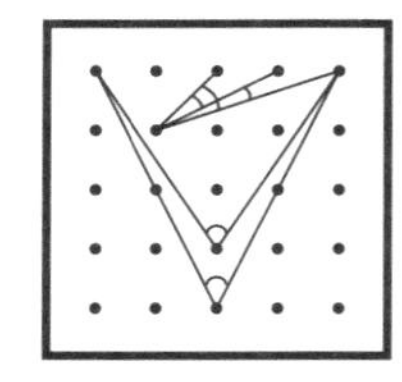
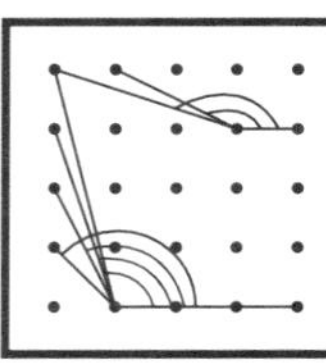
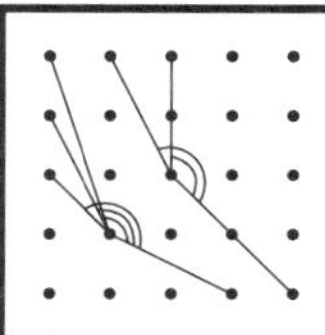
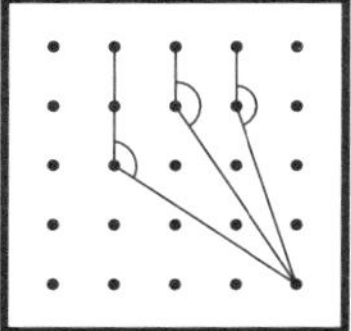

3. Place an index card in the angle. If it matches the corner, it's a right angle.

Page 10:

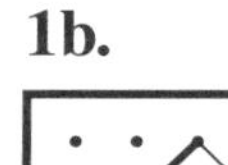

1a. **1b.** **1c.** **2a.** **2b.** **2c.**

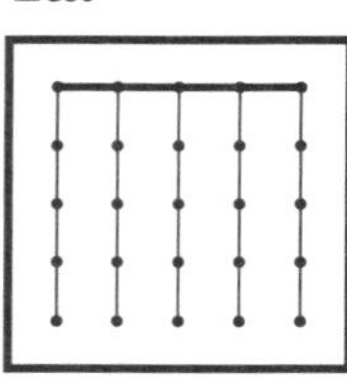

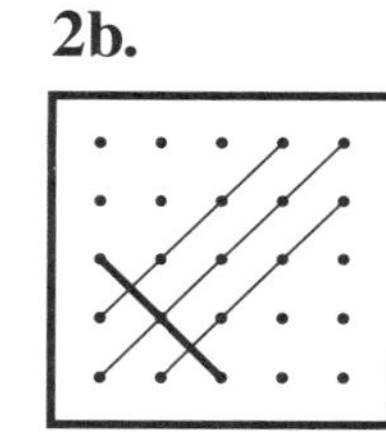
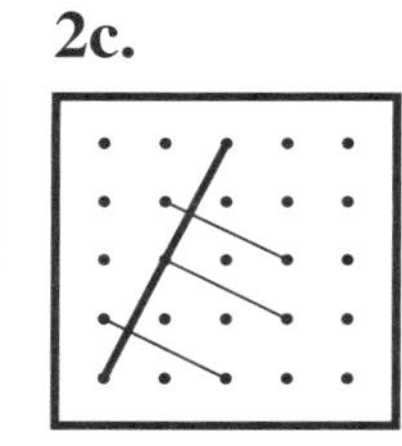

Page 11:

1a. **1b.** **1c.** **2a.** **3.** **4.** Answers vary.

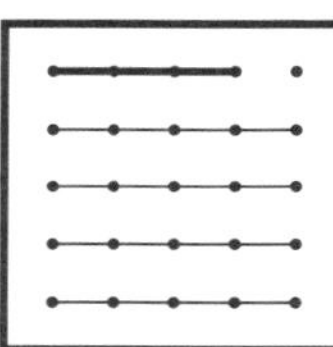
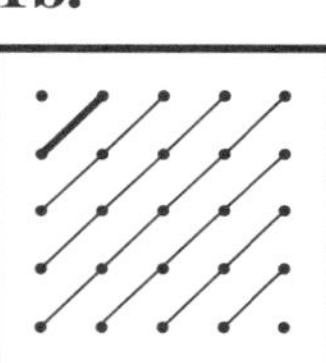
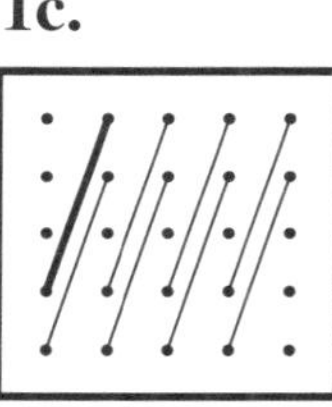
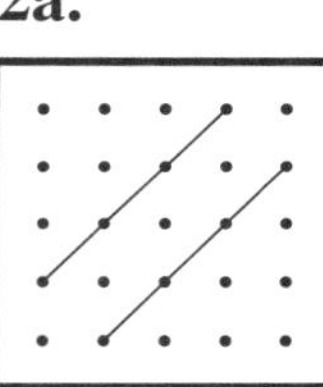
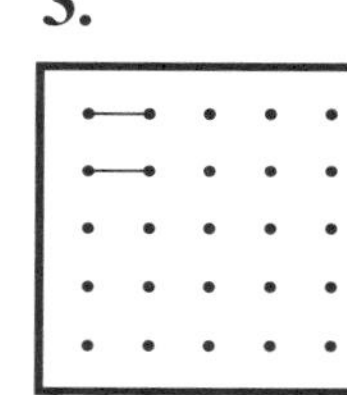

Polygons and Geometric Relationships

Page 14:

1 –2. Answers vary.

Page 15:

Possible answers: **2.** An equilateral triangle cannot be formed on this geoboard.

1. **2.** **3.** **4.** **5.** **6.**

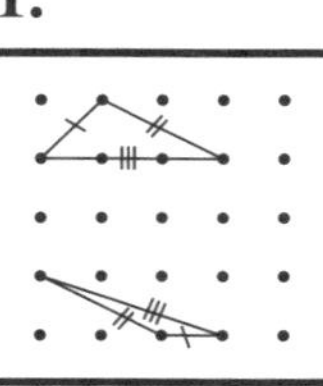
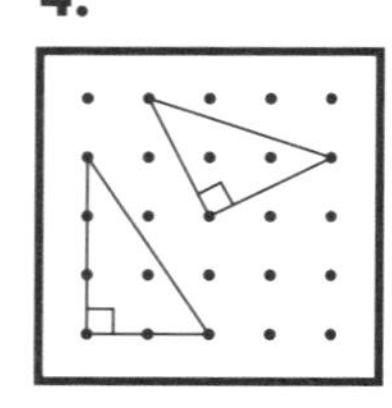
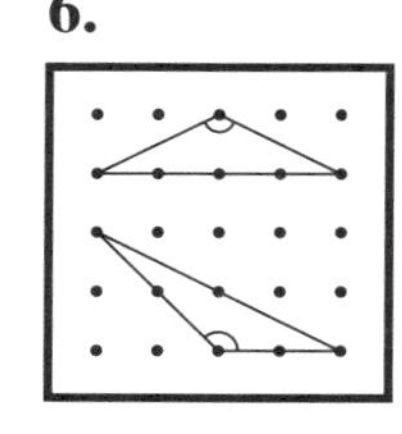

Selected Solutions

Page 16:

Possible answers.

1.

2. 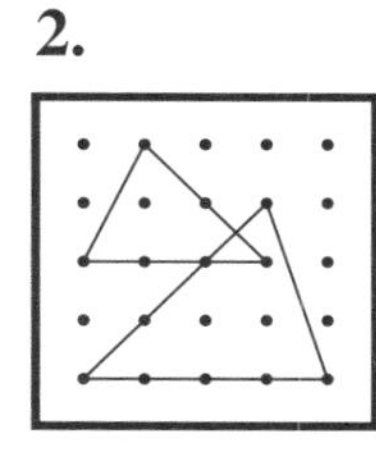

3. 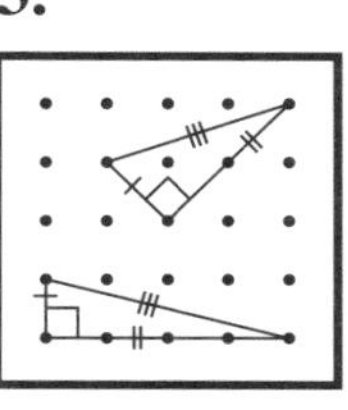

4. 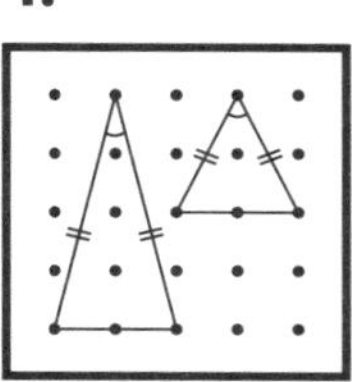

5. 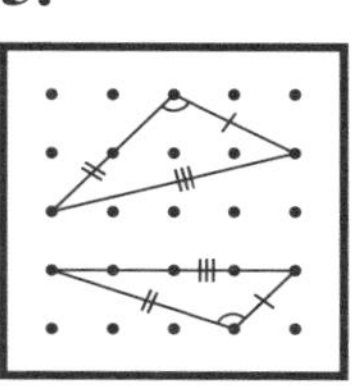

6. 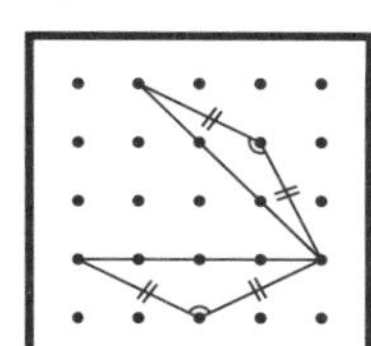

7. scalene right triangle **8.** isosceles right triangle

Page 17:

Possible answers:

1.

2. 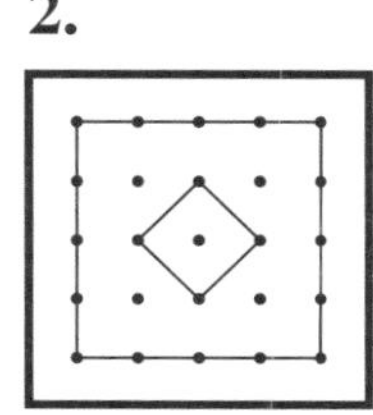

3. 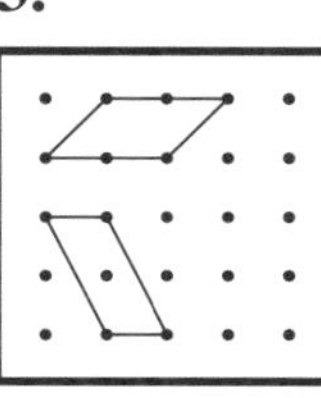

4. 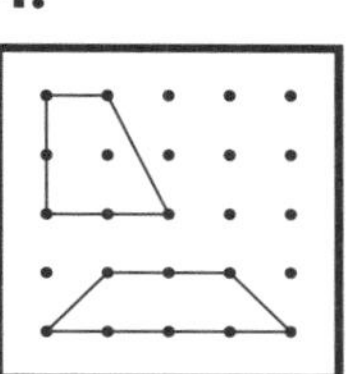

5. 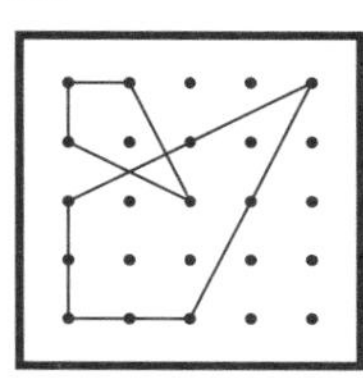

6. 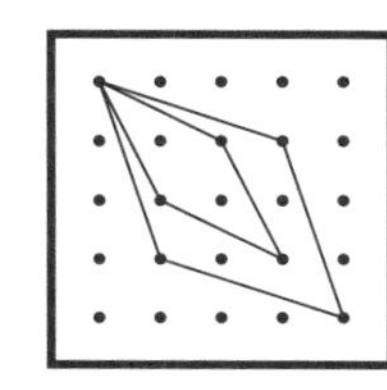

Page 18:

1.

2. 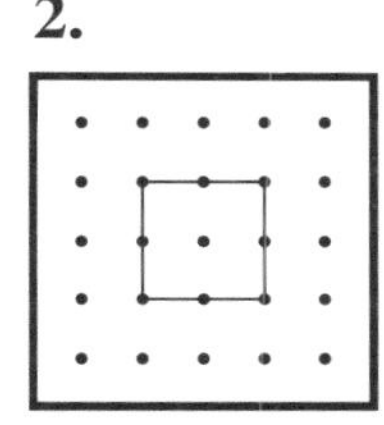

3. 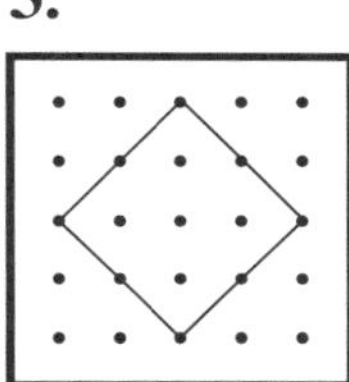

4. 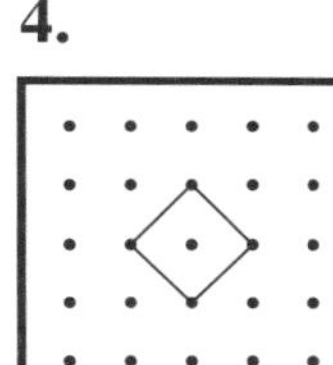

5. 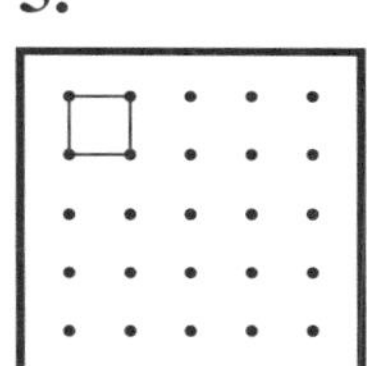

6. 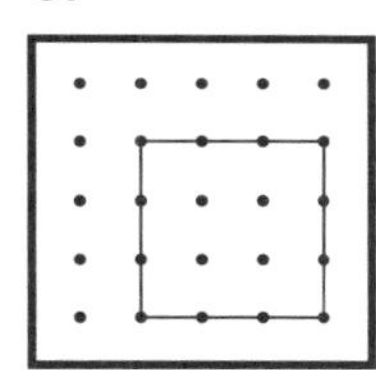

7. 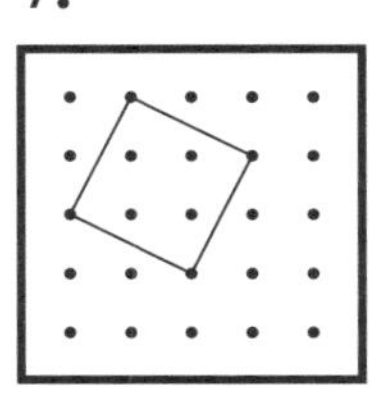

8. 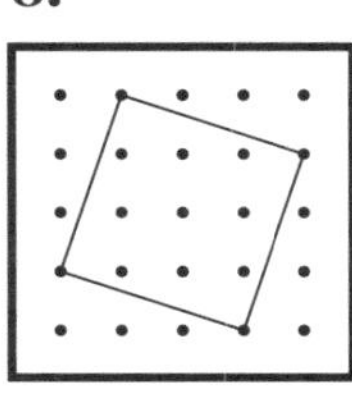

Page 19:

1. trapezoid 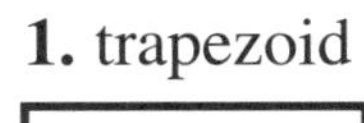

2. rectangle 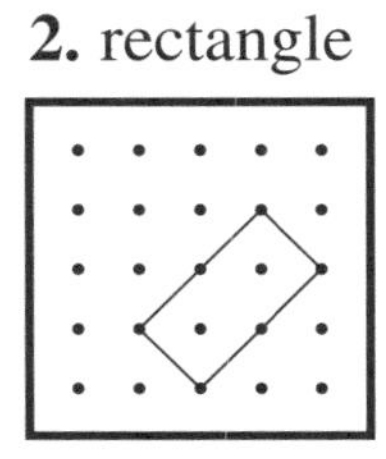

3. square 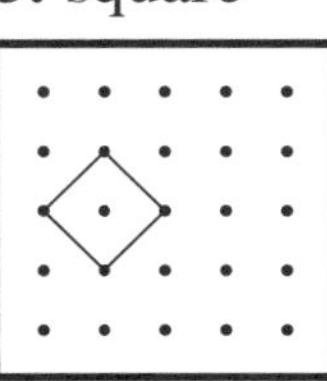

4. trapezoid 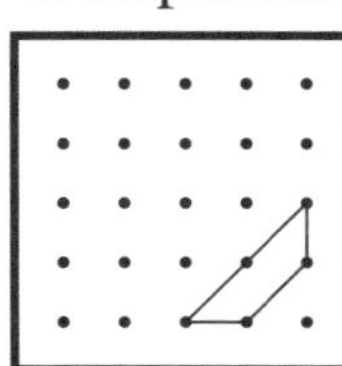

5. kite 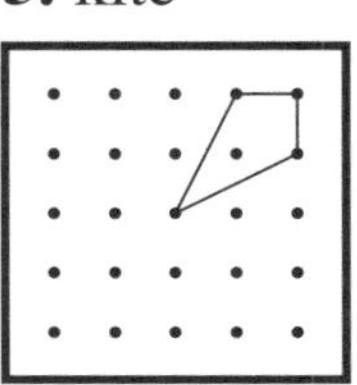

6. parallelogram 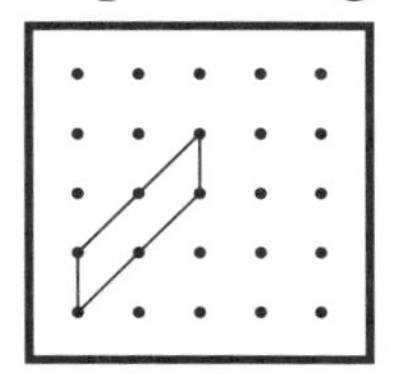

Page 20:

1. 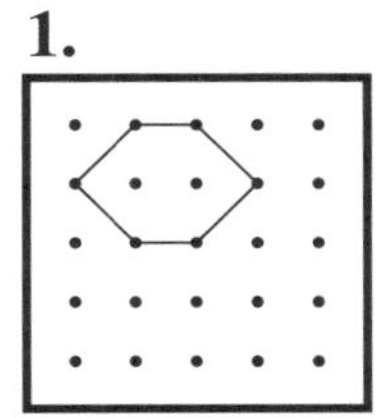

2. 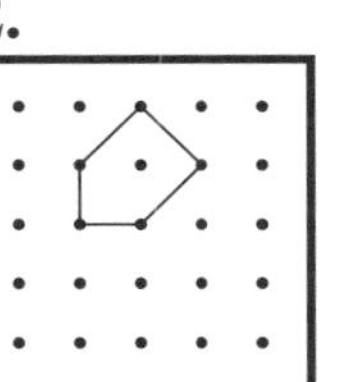

3. 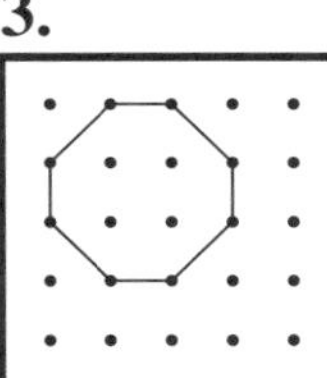

4. 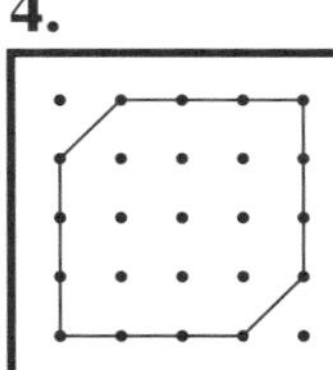

5. 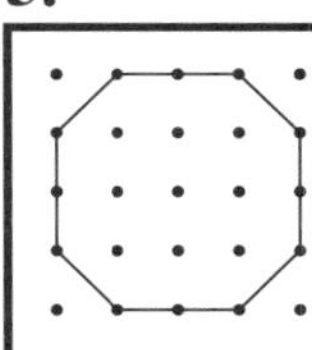

6. 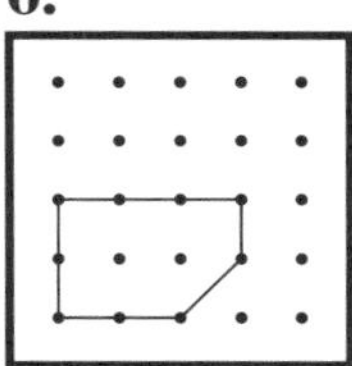

Selected Solutions

Page 20 Continued:

7. Not congruent.
Not similar.

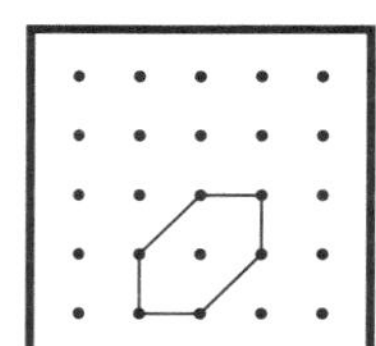

8. Not congruent.
Similar.

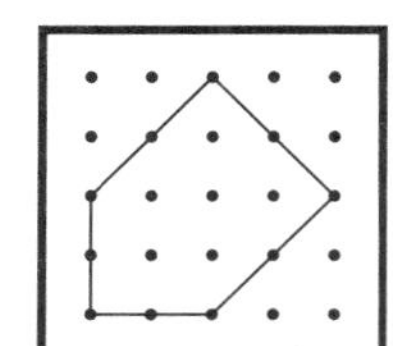

Page 21:

1.

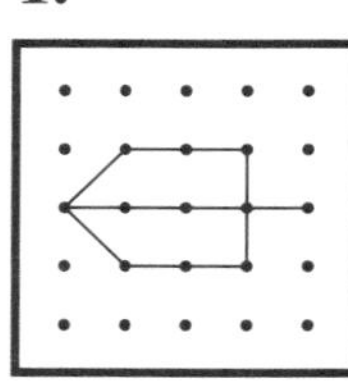

2.

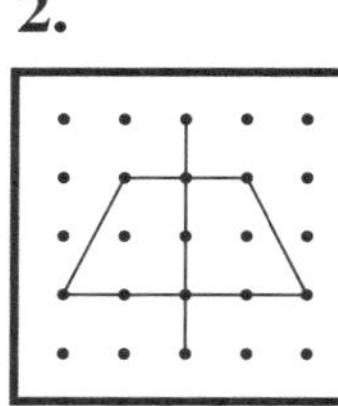

3.

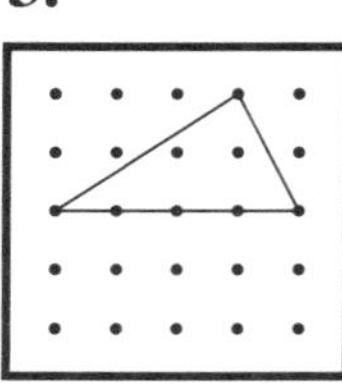

4.

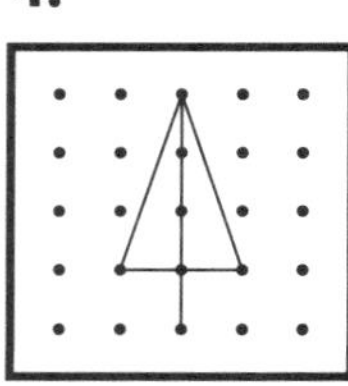

5.

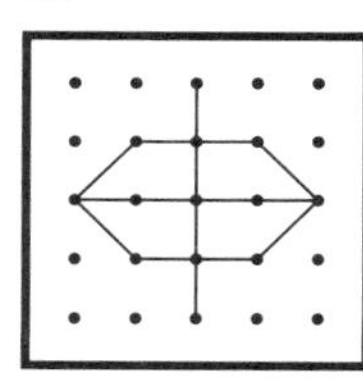

6.

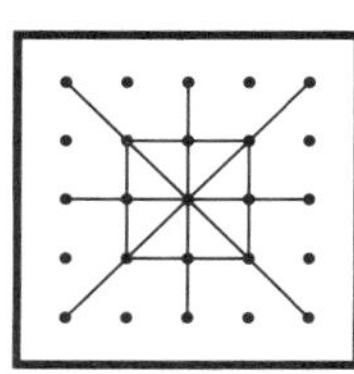

7. isosceles triangle **8.** rectangle
9. trapezoid, parallelogram, irregular quadrilateral
10. square

Page 22:

1. translation

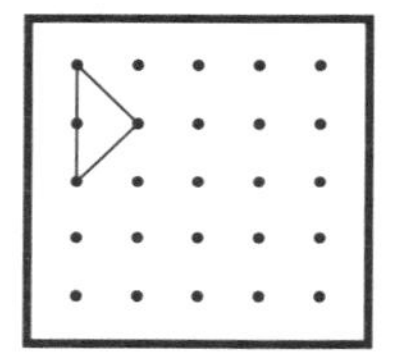

2. reflection

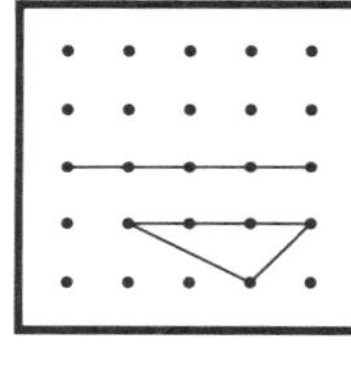

3. rotation

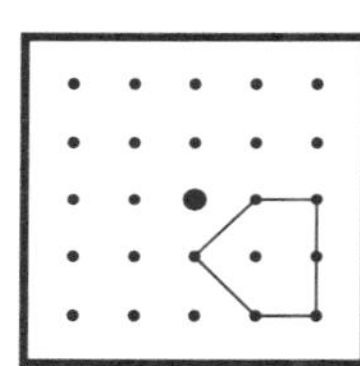

4. ref. & trans.

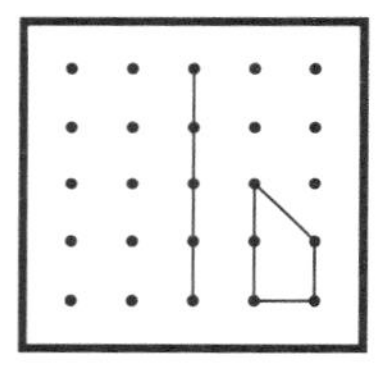

5. ref. & trans.

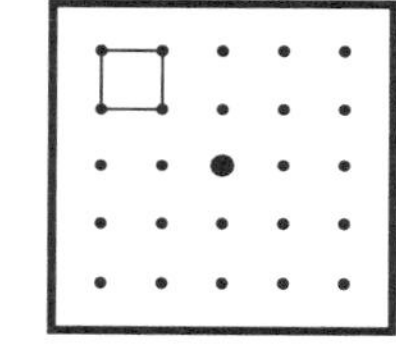

6. trans.

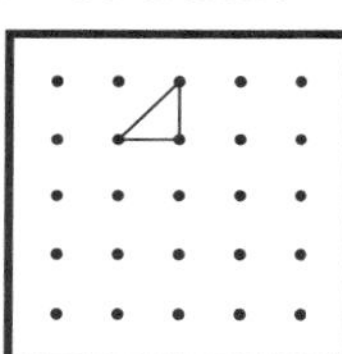

Perimeter and Area

Page 24:

1. 14 **2.** 12 **3.** 14

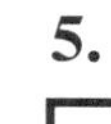

4.

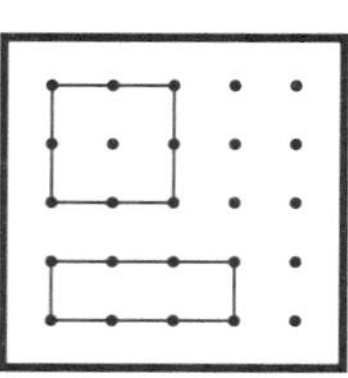

5.

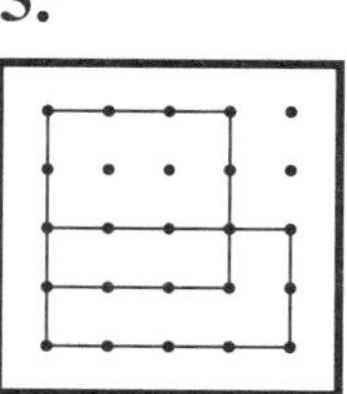

6.

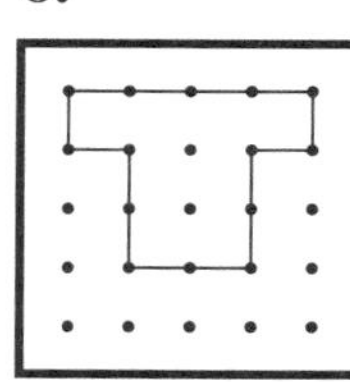

Page 25:

1. 9½ – 10 **2.** 14½–15 **3.** 9½–10 **4.** 11-11½ **5.** 8½–9 **6.** 11–11½

Page 26:

1. 4 **2.** 16 **3.** 6

4.

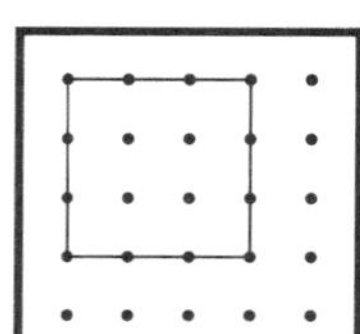

5.

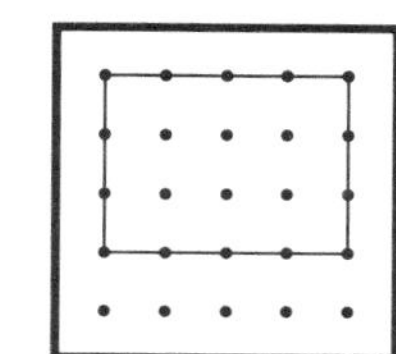

6.

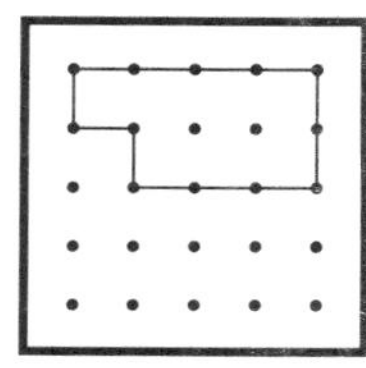

Selected Solutions

Page 27:

1. 6 **2.** 3 **3.** 9 **4.** 6 **5.** 10 **6.** 6

Page 28:

1. $A = 3 \times 3 = 9$ **2.** $A = 2 \times 4 = 8$ **3.** $A = 3 \times 4 = 12$

4. $A = \frac{1}{2} \times 3 \times 3 = 4\frac{1}{2}$ **5.** $A = \frac{1}{2} \times 2 \times 4 = 4$ **6.** $A = \frac{1}{2} \times 4 \times 4 = 8$

Fractions

Page 30:

1. $\frac{1}{2}$

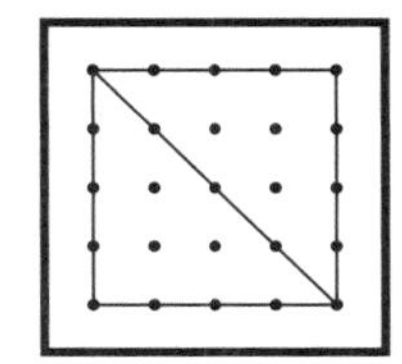
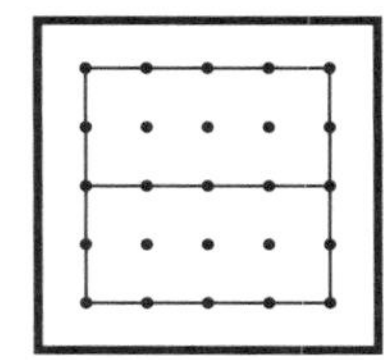
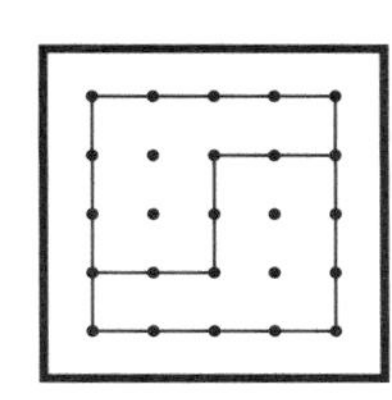

2. $\frac{1}{4}$

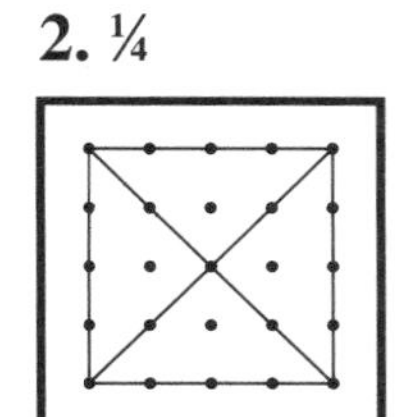
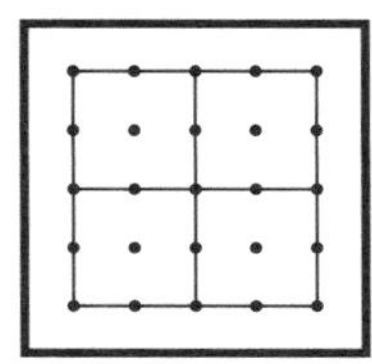
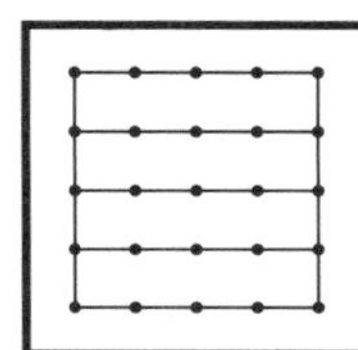

3. $\frac{1}{8}$

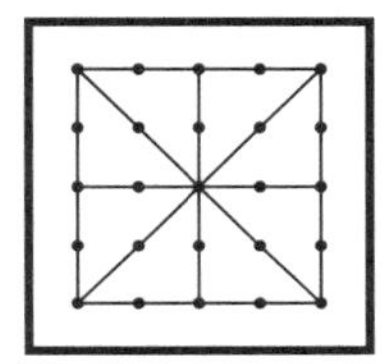
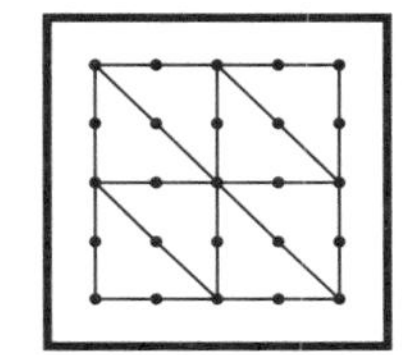
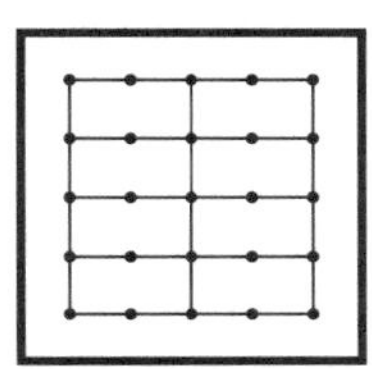

Page 31:

1.

2.

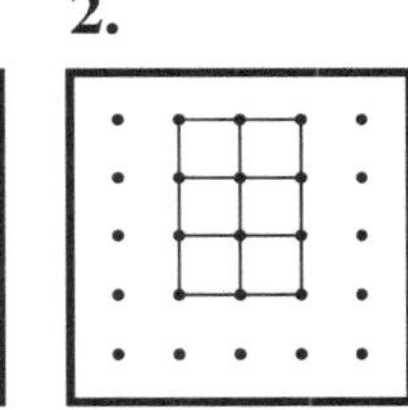

3.

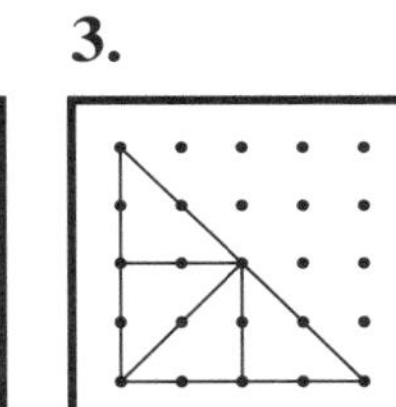

4.

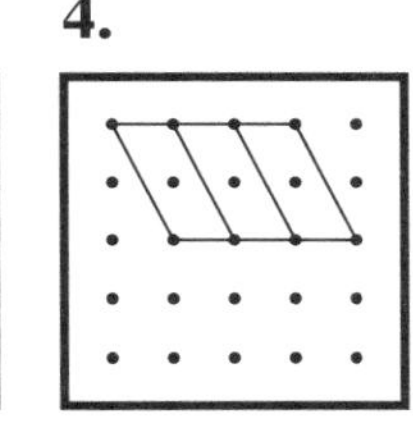

5.

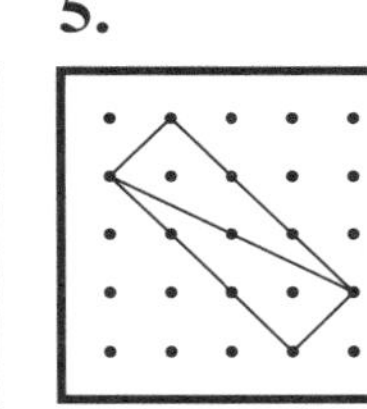

6.

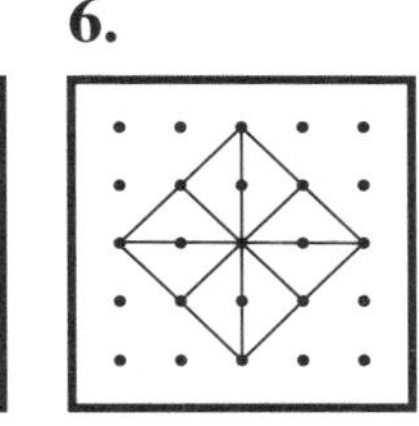

7.

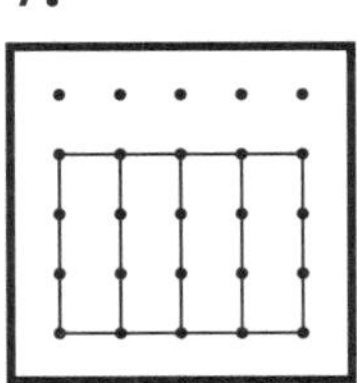

8.

9.

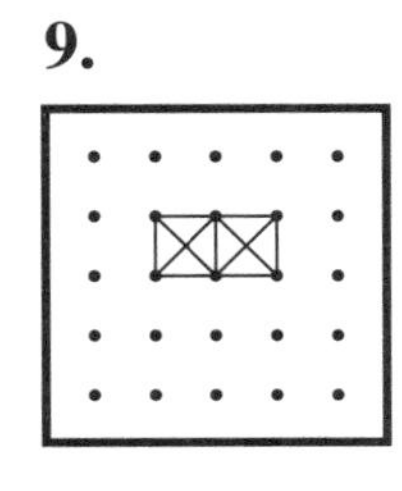

Page 32:

1.

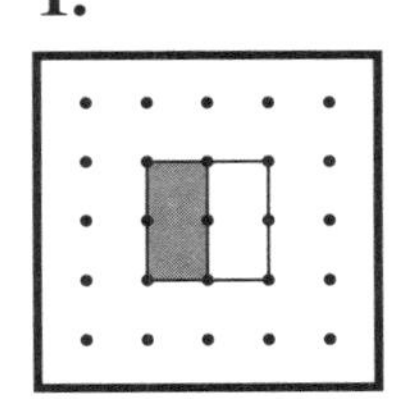

2.

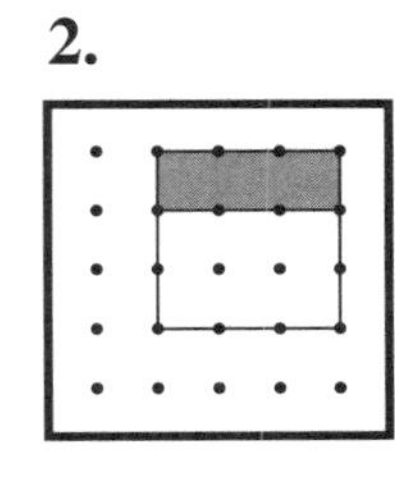

3.

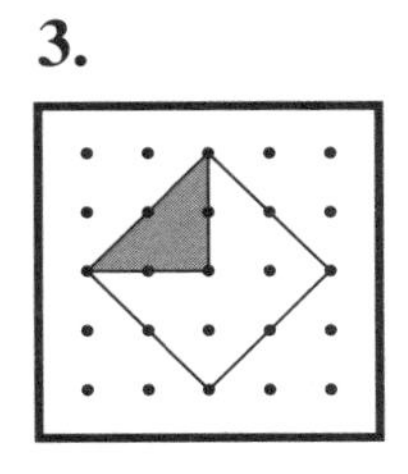

4.

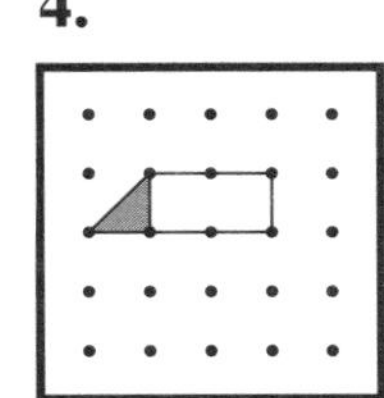

5.

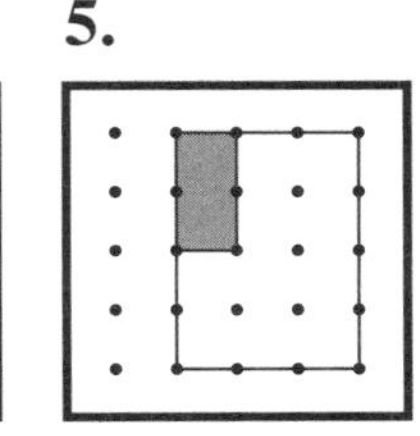

6.

Selected Solutions

Page 32 Continued:

7.

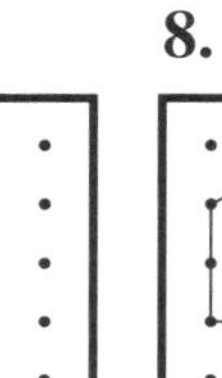

8.

9.

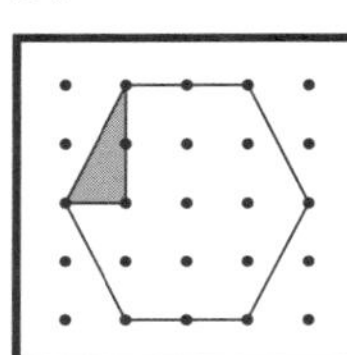

Page 33:

1.

2.

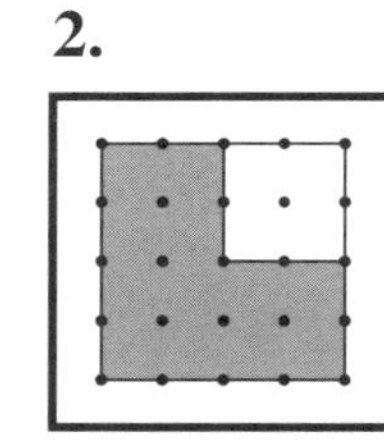

3.

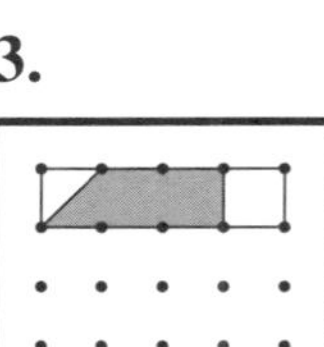

4.

5.

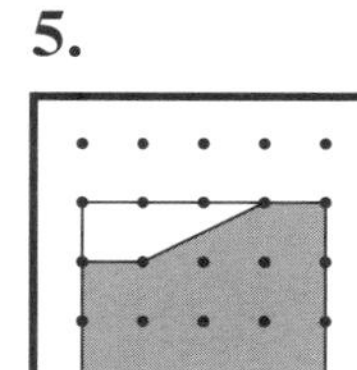

6.

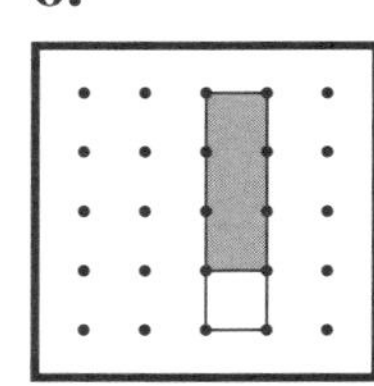

7.

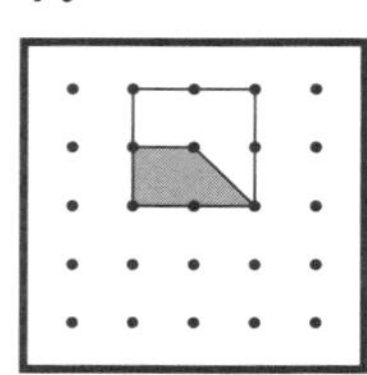

8.

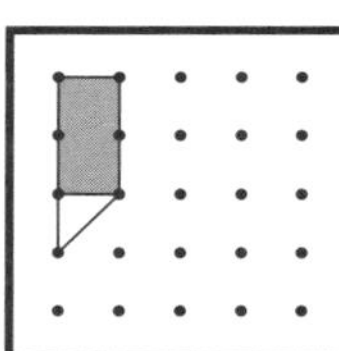

9.

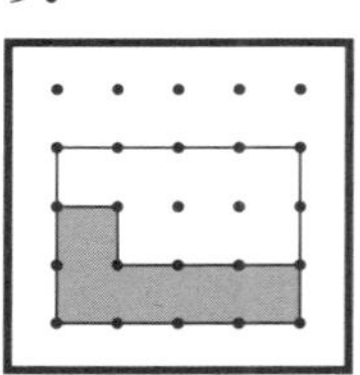

Page 34:

2.

3.

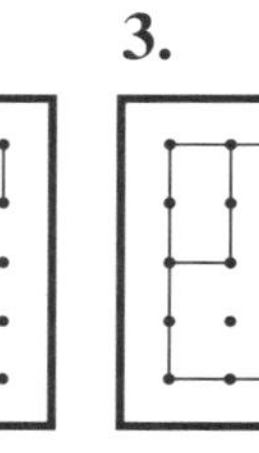

4.

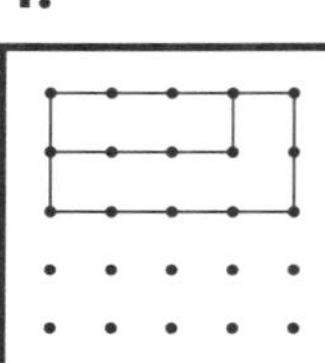

5.

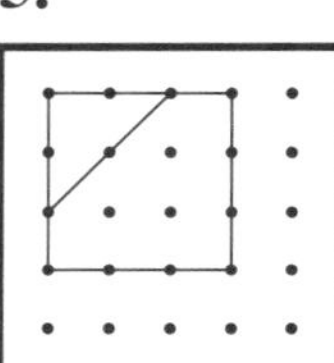

6.

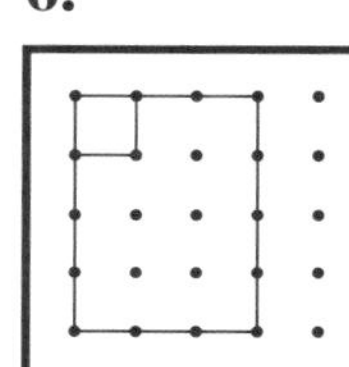

7.

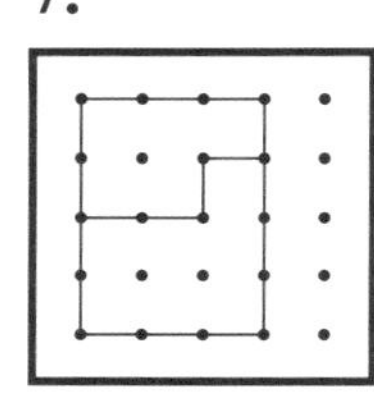

8.

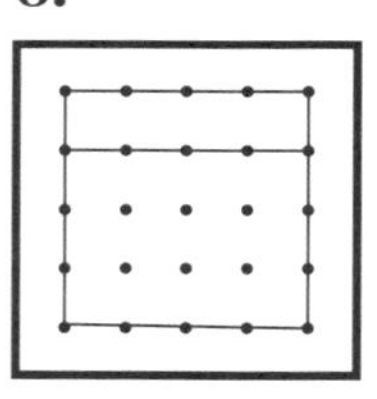

9.

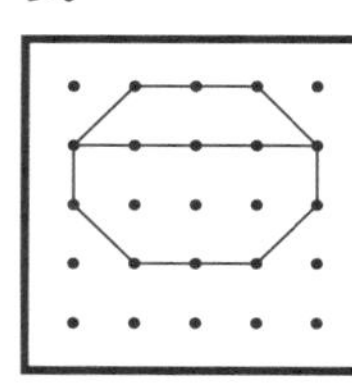

Parts of a Circle

Page 36

1–3.

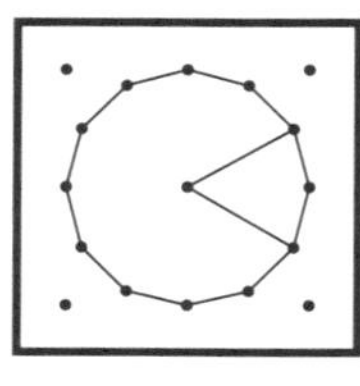

4.

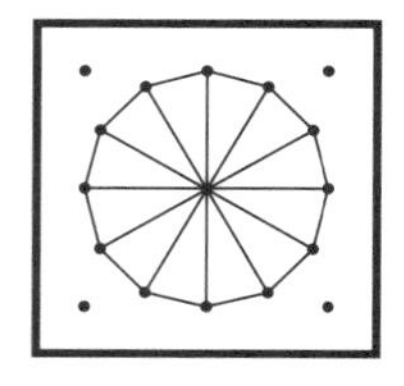

5. Yes, all the points on the circle are the same distance from the center.

Selected Solutions

Page 37

1.

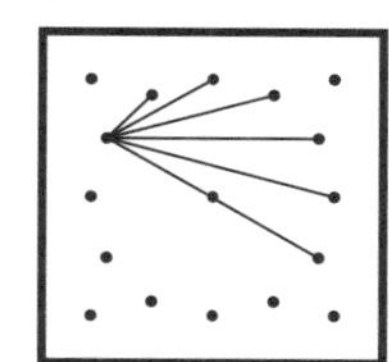

2. Yes; the longest chord. **3.** 6
4. Diameter; it's made of two radii

Page 38:

1–5. Possible answers.

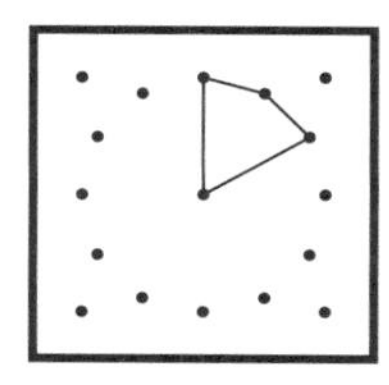
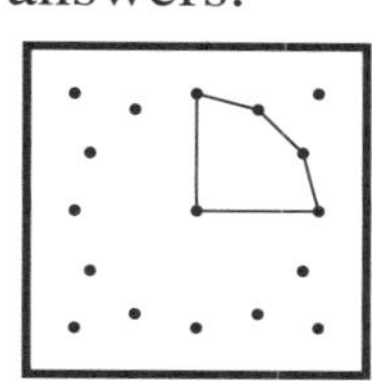
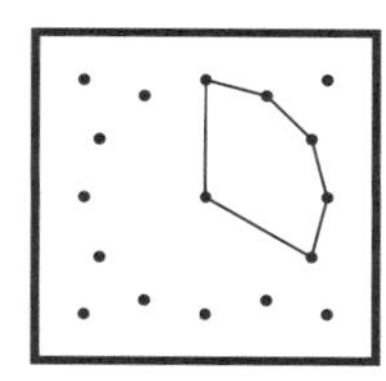
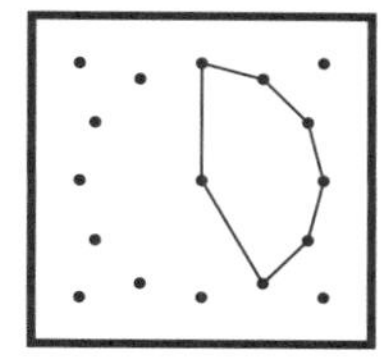
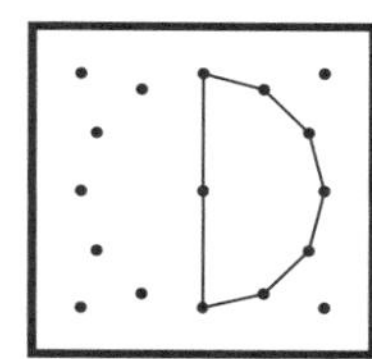

Page 39:

3. About 3 times. **4.** A little more than 3 times.
6. The answer for each should be close to 3.14.

Fractions and Angle Measurements

Page 41:

1.

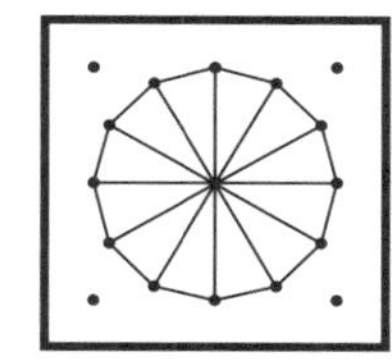

2. 12 **3.** $\frac{1}{12}$ **4.** $\frac{6}{12} = \frac{1}{2}$, $\frac{3}{12} = \frac{1}{4}$
5. For every pin included to form a central angle, the fraction increases by $\frac{1}{12}$.

6.

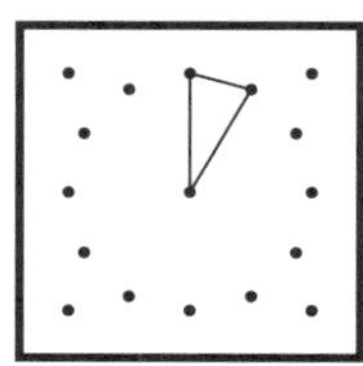

7.

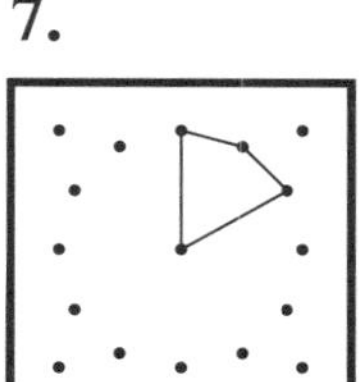

8.

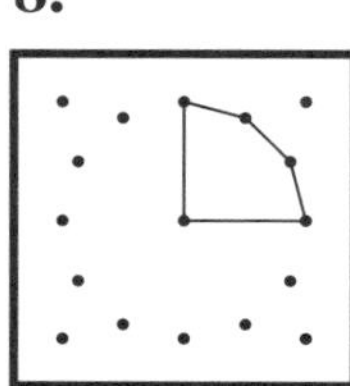

9.

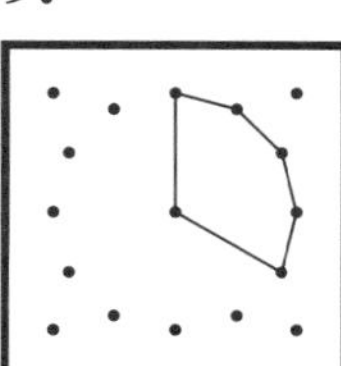

10.

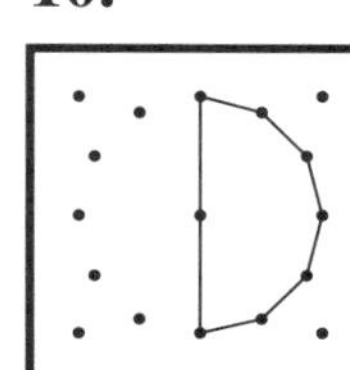

Page 42:

1.

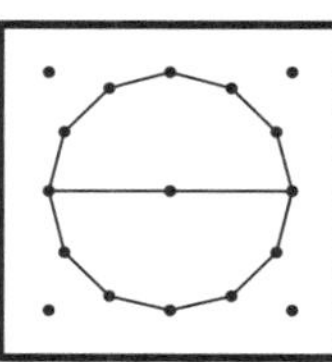

360° ÷ 2 = 180°

2.

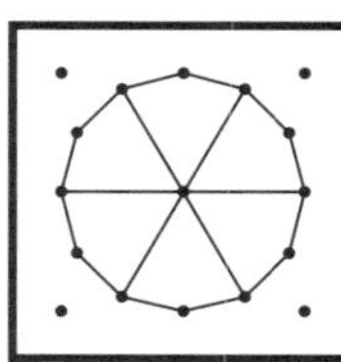

360° ÷ 6 = 60°

3.

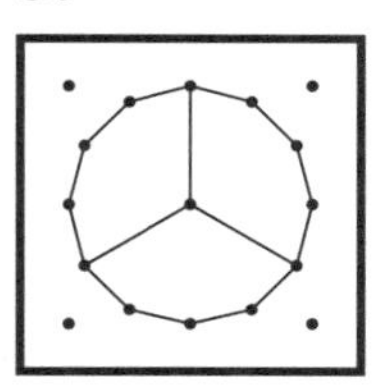

360° ÷ 3 = 120°

4.

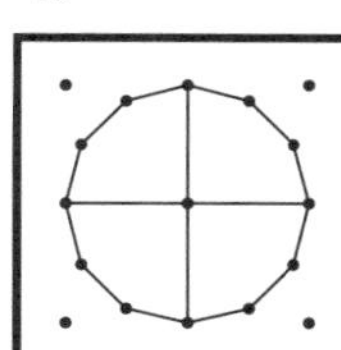

360° ÷ 4 = 90°

5.

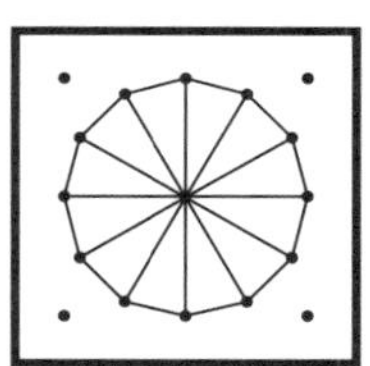

360° ÷ 12 = 30°

6.

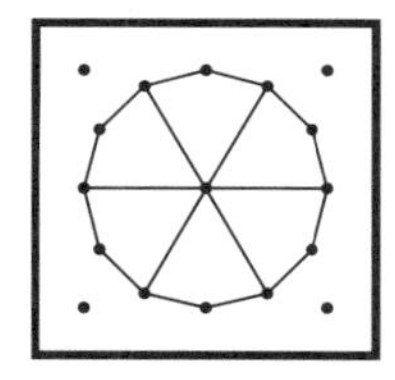

360° ÷ 6 = 60°

Selected Solutions

Page 43:

1. **2.** **3.** **4.** **5.** **6.**

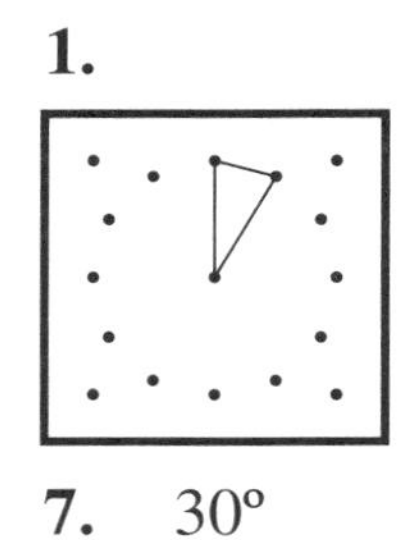
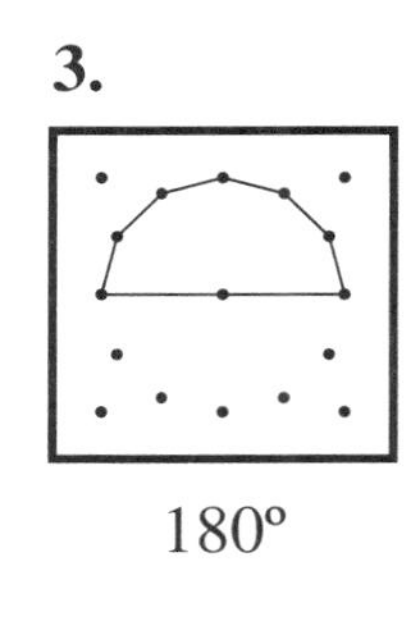
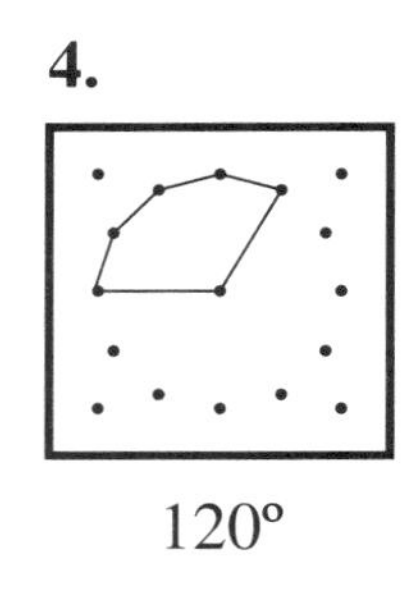
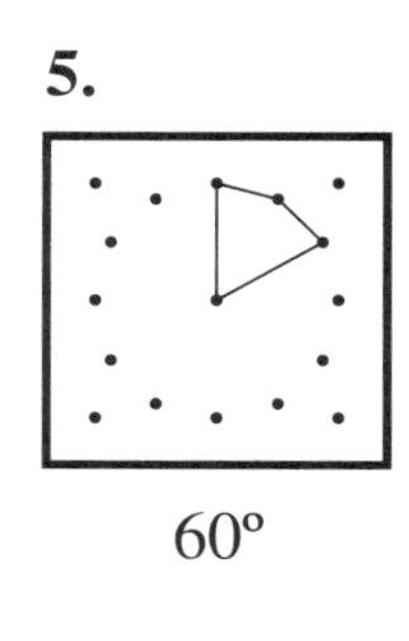
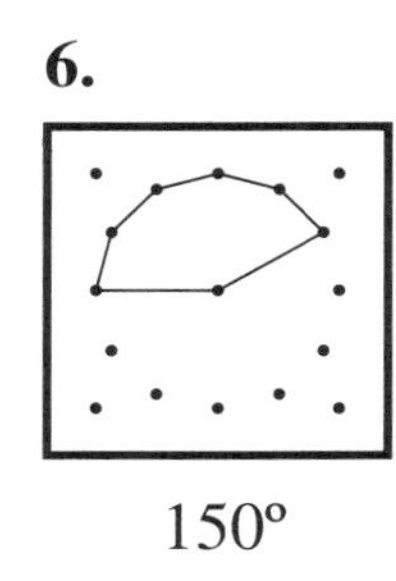

7. 30º 90º 180º 120º 60º 150º

Page 44:

1. **2.** **3.** **4.** **5.**

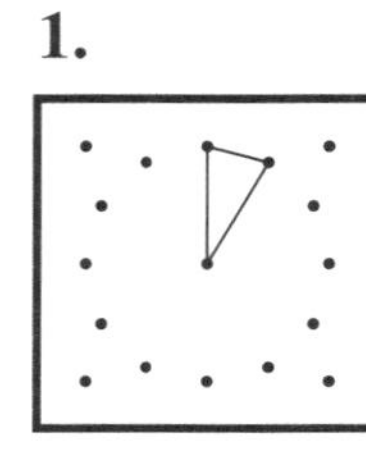
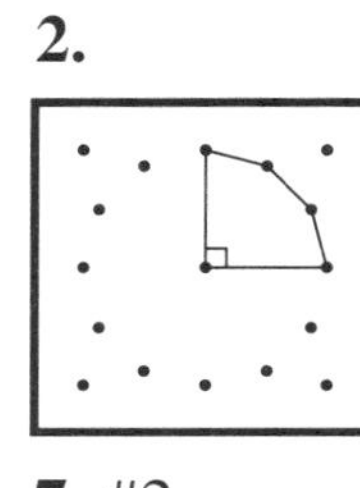
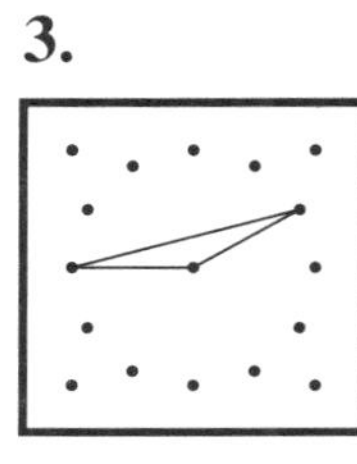
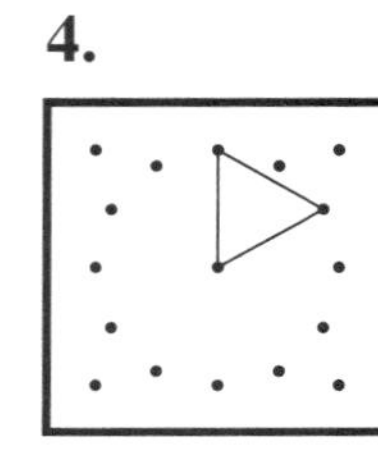
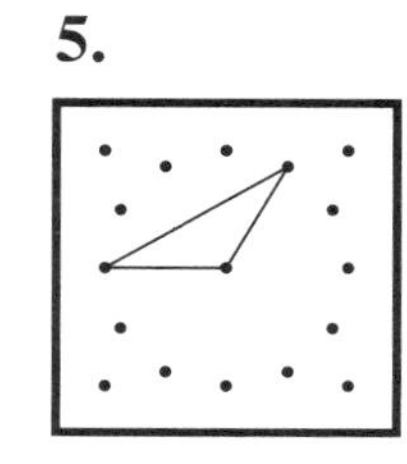

6. #1, #4 **7.** #2 **8.** #3, #5 **9.** #4 **10.** all **11.** none

Page 46: Possible answers for 1–8.

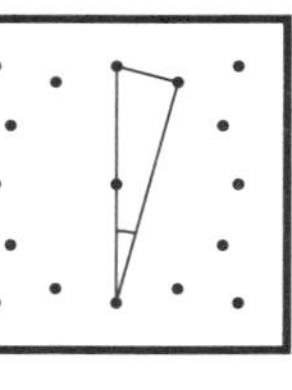
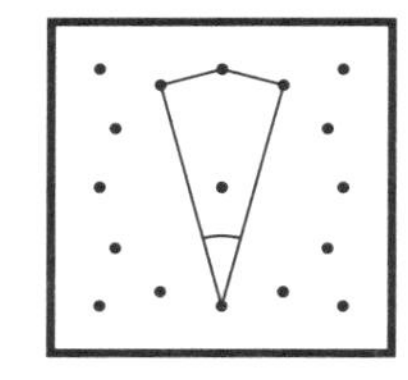
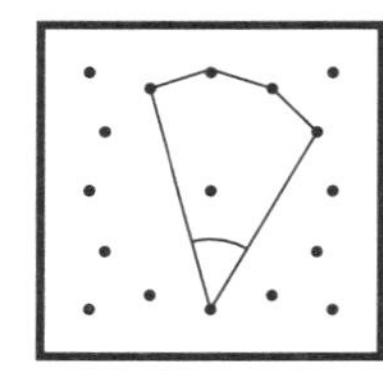
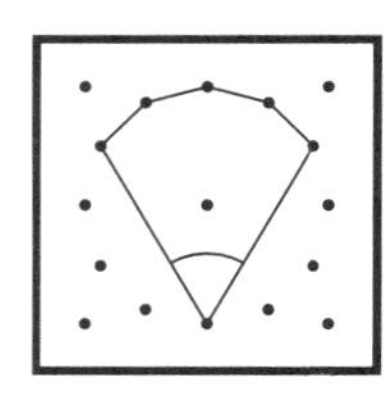
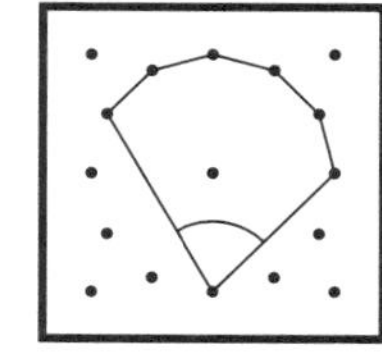
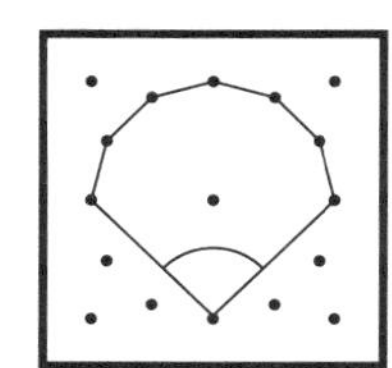
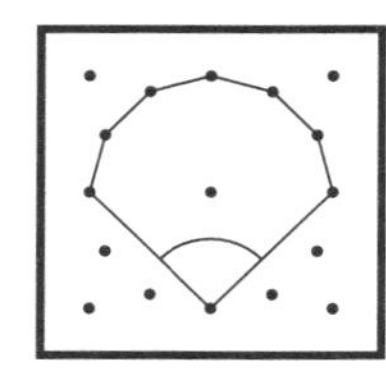
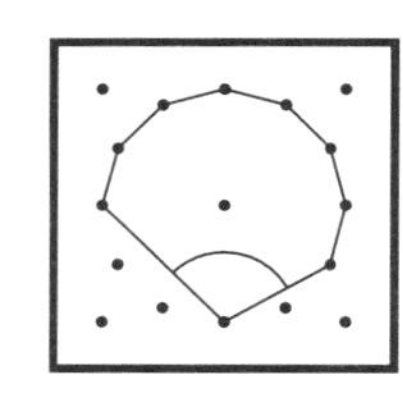
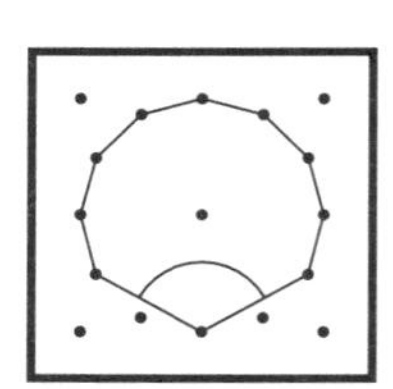

Page 47:

1. 60º, 30º **2.** 90º, 45º **3.** 30º, 15º **4.** 150º, 75º **5.** 180º, 90º **6.** 120º, 60º

Page 48:

1. **2.** **3.** **4.** **5.** **6.**

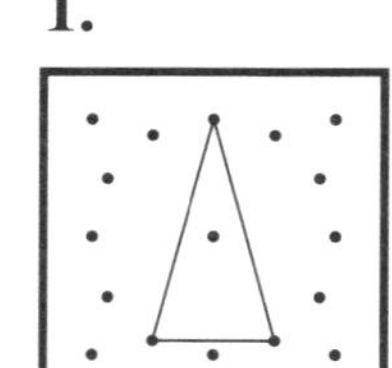
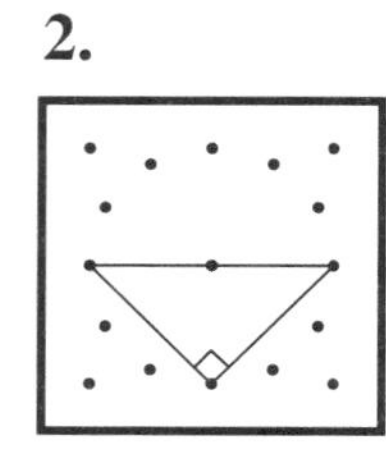
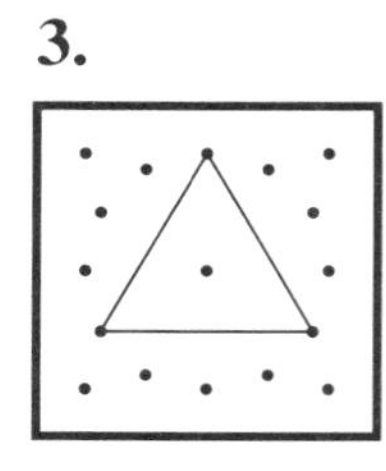
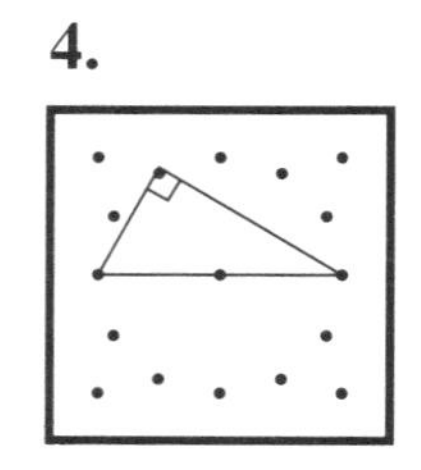
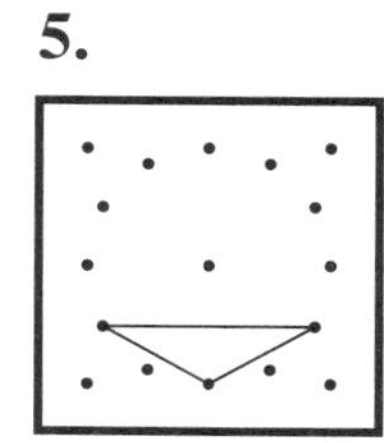
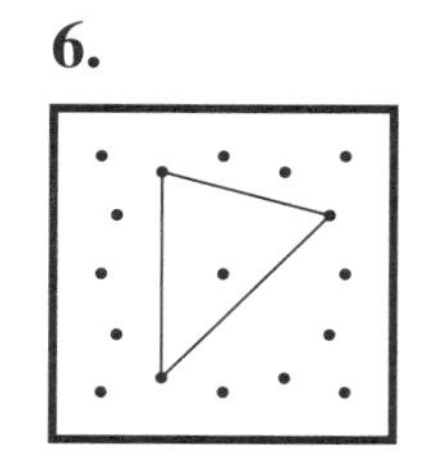

Selected Solutions

Page 49:

1. 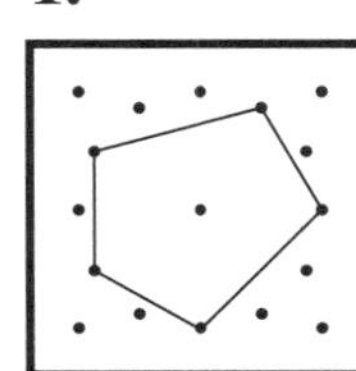
2. 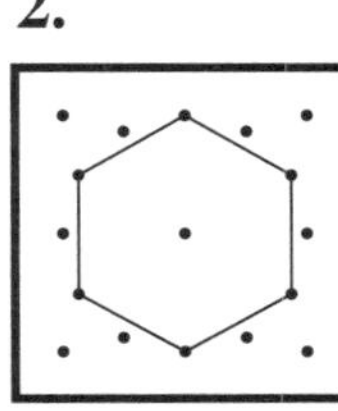
3. 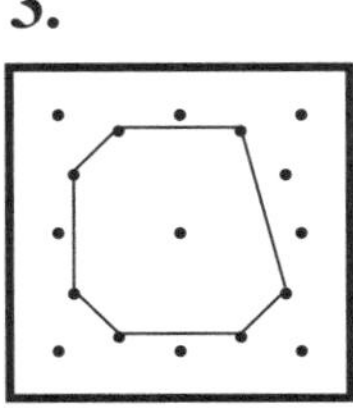
4. 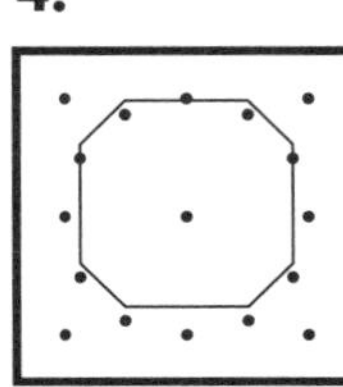
5. 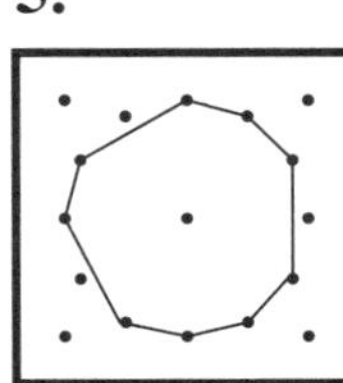
6. 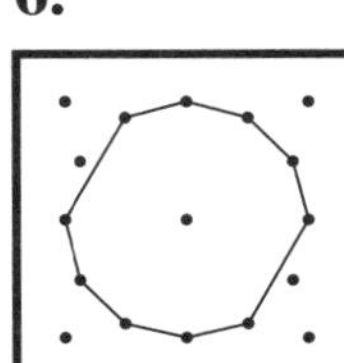

7. #1: pentagon; #2: hexagon; #4: octagon
(#3: heptagon; #5: nonagon; #6: decagon)

Add Fractions

Page 51:

1. 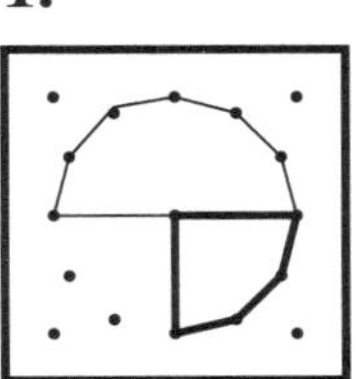
$\frac{1}{2} + \frac{1}{4} =$
$\frac{6}{12} + \frac{3}{12} = \frac{9}{12} = \frac{3}{4}$

2. 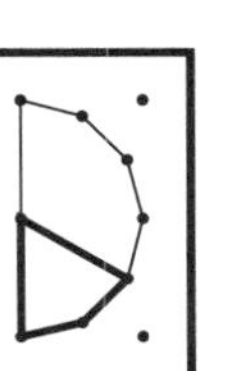
$\frac{1}{3} + \frac{1}{6} =$
$\frac{4}{12} + \frac{2}{12} = \frac{6}{12} = \frac{1}{2}$

3. 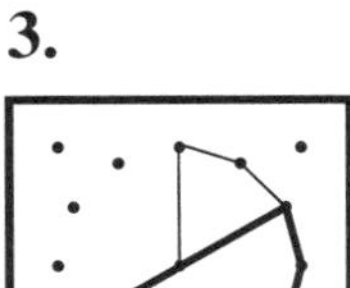
$\frac{1}{6} + \frac{1}{2} =$
$\frac{2}{12} + \frac{6}{12} = \frac{8}{12} = \frac{2}{3}$

4. 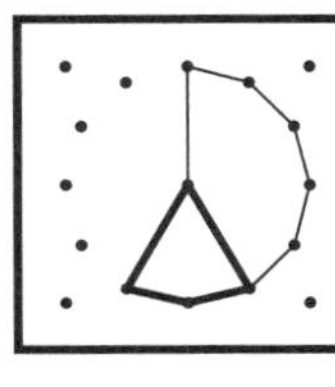
$\frac{5}{12} + \frac{1}{6} =$
$\frac{5}{12} + \frac{2}{12} = \frac{7}{12}$

5. 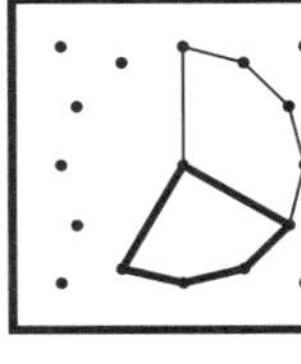
$\frac{1}{3} + \frac{1}{4} =$
$\frac{4}{12} + \frac{3}{12} = \frac{7}{12}$

6. 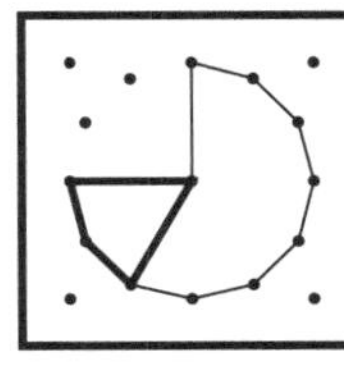
$\frac{7}{12} + \frac{1}{6} =$
$\frac{7}{12} + \frac{2}{12} = \frac{9}{12}$

7. The pins count off $\frac{1}{12}$s.

8. Show the first fraction and then count on by $\frac{1}{12}$s for the second fraction.

Page 52:

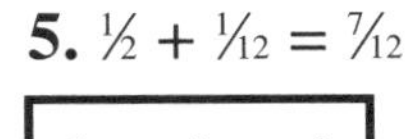

1. $\frac{1}{12} + \frac{1}{6} = \frac{1}{4}$ **2.** $\frac{1}{3} + \frac{1}{6} = \frac{1}{2}$ **3.** $\frac{1}{6} + \frac{1}{4} = \frac{5}{12}$ **4.** $\frac{1}{3} + \frac{1}{2} = \frac{5}{6}$ **5.** $\frac{1}{2} + \frac{1}{12} = \frac{7}{12}$

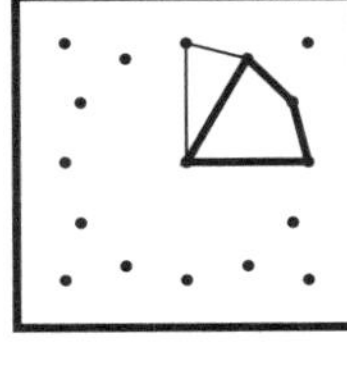
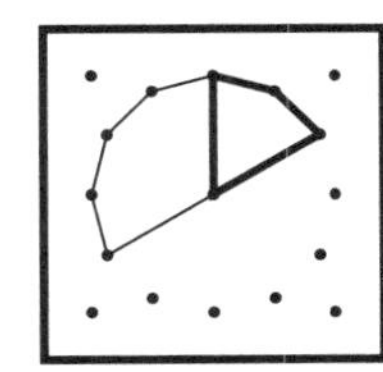
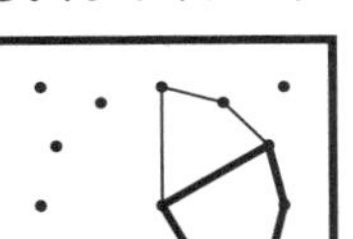
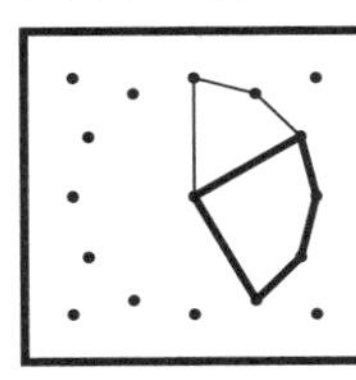
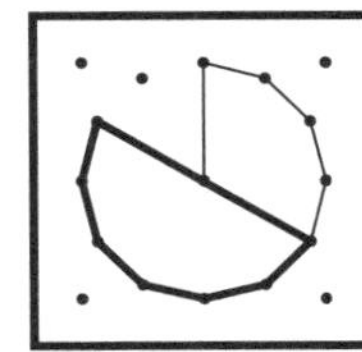
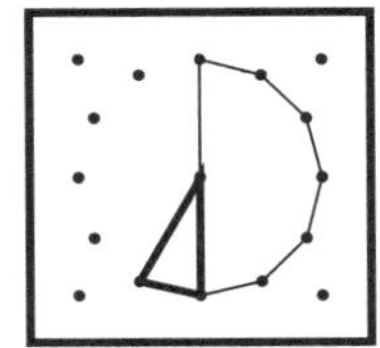

6. $\frac{1}{2} + \frac{1}{3} = \frac{5}{6}$ **7.** $\frac{1}{2} + \frac{1}{2} = 1$ **8.** $\frac{1}{3} + \frac{1}{6} + \frac{1}{12} = \frac{7}{12}$

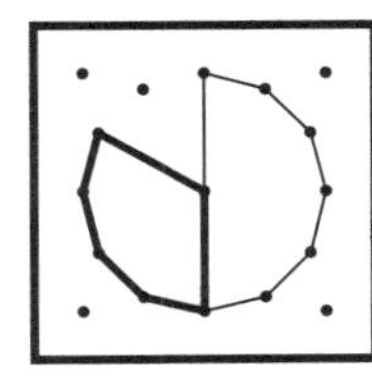
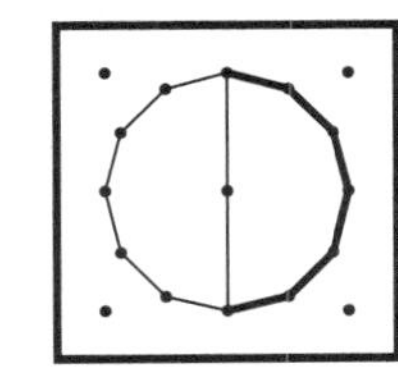
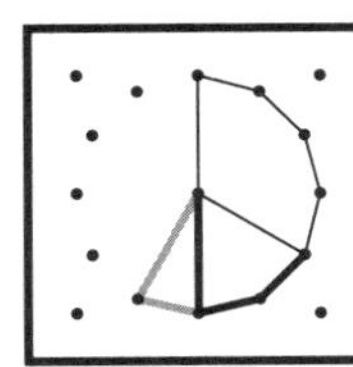

Page 53:

1. $\frac{1}{3} + \frac{1}{3} = \frac{2}{3}$
($\frac{4}{12} + \frac{4}{12} = \frac{8}{12}$)

2. $\frac{1}{12} + \frac{1}{12} = \frac{2}{12} = \frac{1}{6}$

3. $\frac{1}{6} + \frac{1}{6} = \frac{2}{6} = \frac{1}{3}$

4. $\frac{1}{6} + \frac{1}{3} = \frac{1}{2}$
($\frac{2}{12} + \frac{4}{12} = \frac{6}{12}$)

5. $\frac{5}{12} + \frac{1}{12} = \frac{6}{12} = \frac{1}{2}$

6. $\frac{1}{6} + \frac{1}{2} = \frac{2}{3}$
($\frac{2}{12} + \frac{6}{12} = \frac{8}{12}$)

Glossary

Arc A part of the circle cut off by the sides of a central angle.

Angle The union of two rays with a common endpoint.

Acute angle	measures between 0 and 90 degrees
Right angle	measures 90 degrees
Obtuse angle	measures between 90 and 180 degrees
Straight angle	measures 180 degrees

Area The amount of surface covered by a figure; measured in square units.

Central angle An angle formed by two radii of the circle. The vertex is the center of the circle.

Central angle triangle A triangle formed with the central angle of a circle and a chord connecting the two endpoints of the central angle on the circle.

Chord A line segment connecting two points on a circle.

Circle A simple closed curve in a plane. Every point on the circle is an equal distance from the center.

Circumference The distance around a circle. It is a little more than 3, or about 3.14, times the length of its diameter.

Congruent figures Figures having the same shape and size.

Degree A unit of measure of an angle. There are 360 degrees in a complete rotation.

Diameter A chord that passes through the center of the circle.

Equivalent fractions A fractional value expressed by two or more names. Fractions showing the same value.

Fraction A part of a whole region or set.

- **Proper fraction** A fraction whose value is between 0 and 1.
- **Unit fraction** A fraction whose numerator is 1; represents one part of a whole region or set.
- **Lowest-terms fraction** A fraction having 1 as the greatest common factor of its numerator and denominator.

Inscribed angle An angle whose vertex is on the circle. Each side of the angle is a chord. Its size is one-half the size of the central angle cutting off the same arc.

Inscribed polygon A polygon having all vertices on the circle. Each side of the polygon is a chord.

Line segment A part of a line having two endpoints.

Parallel lines Lines in the same plane that do not intersect.

Perimeter The distance around a figure; measured in linear units.

Perpendicular lines Lines at or forming right angles.

Polygon A simple closed figure in a plane made up of straight line segments. A ***regular polygon*** has congruent sides.

Triangle	3 sides	**Heptagon**	7 sides
Quadrilateral	4 sides	**Octagon**	8 sides
Pentagon	5 sides	**Nonagon**	9 sides
Hexagon	6 sides	**Decagon**	10 sides

Quadrilateral A polygon with 4 sides.

Kite	2 pairs of adjacent congruent sides
Trapezoid	1 pair of opposite sides parallel
Parallelogram	2 pairs of opposite sides parallel
Rectangle	2 pairs of opposite sides parallel, right angles
Rhombus	2 pairs of opposite sides parallel, congruent sides
Square	2 pairs of opposite sides parallel, congruent sides, right angles

Radius The distance from the center of the circle to a point on the circle.

Reflection (flip) A motion of a figure about a line; mirror image.

Rotation (turn) A motion of a figure around a fixed point by a certain amount of degrees in a certain direction.

Similar figures Figures having the same shape and corresponding proportional sides; an enlarged or reduced form of a figure.

Symmetric figure A figure identical to its own reflection about a line of symmetry.

Translation (slide) A motion of a figure involving direction and distance along a straight line.

Triangle A polygon with 3 sides.

Acute triangle	all angles are acute
Right triangle	contains a right angle
Obtuse triangle	contains an obtuse angle
Equilateral triangle	all sides are congruent
Isosceles triangle	at least two sides are congruent
Scalene triangle	all sides are unequal to each other

Vertex The common endpoint of two sides that form an angle.

5 x 5 Geoboard Recording Sheet

Name........................

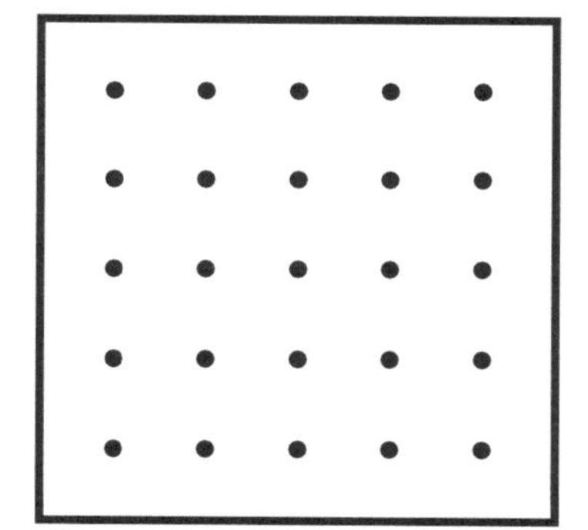
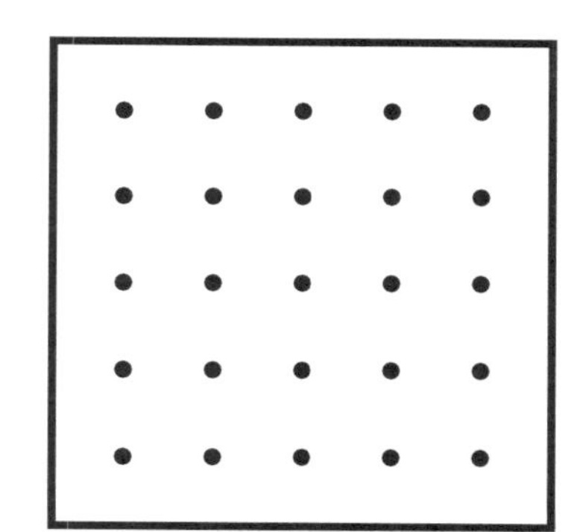
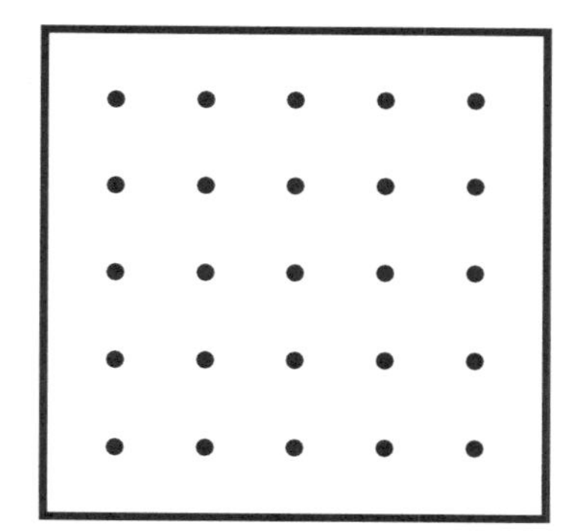
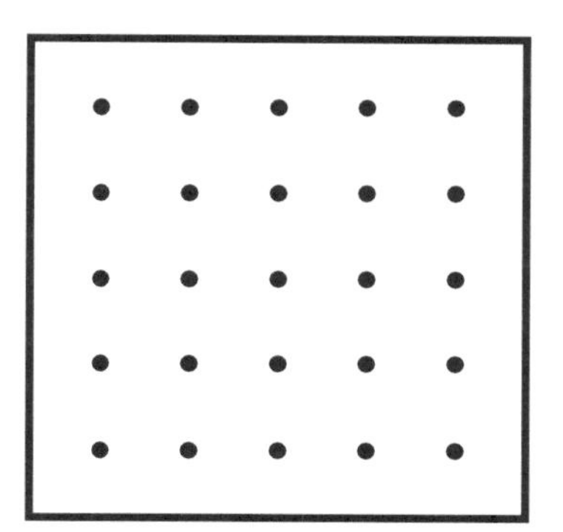
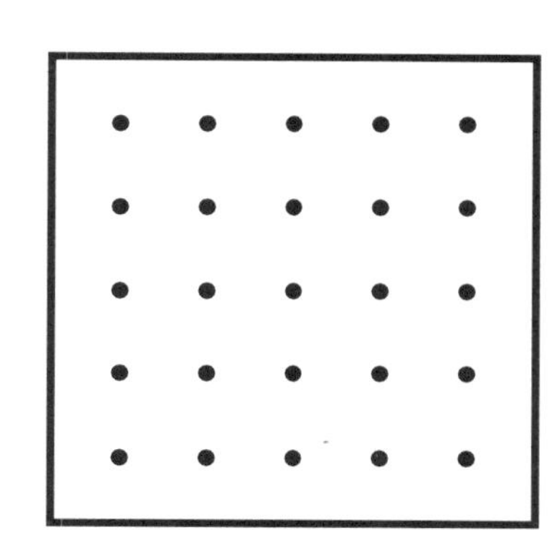
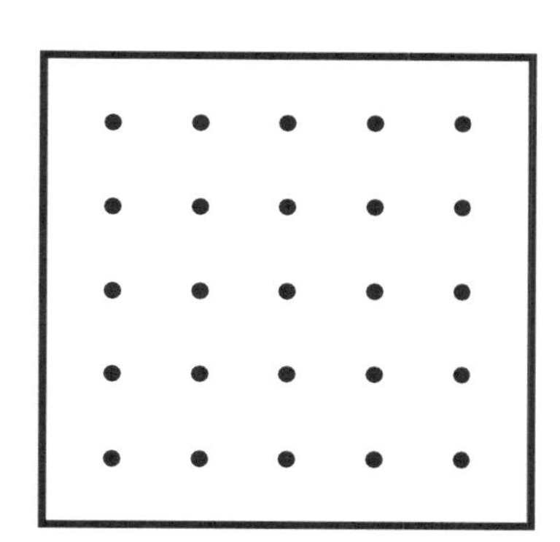
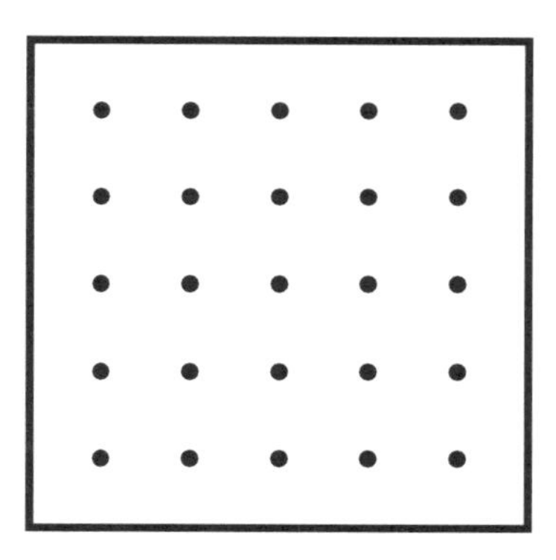
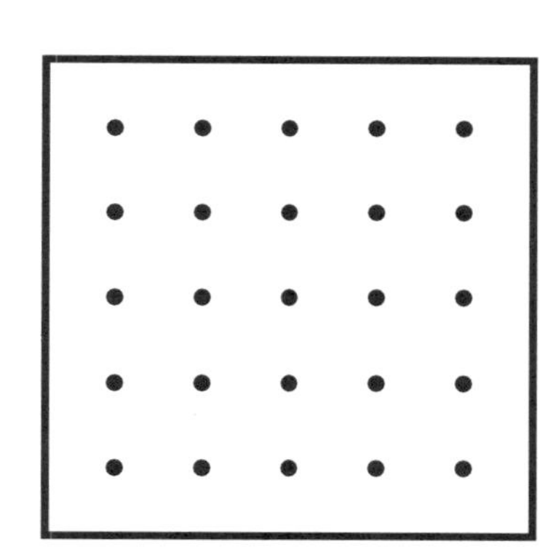
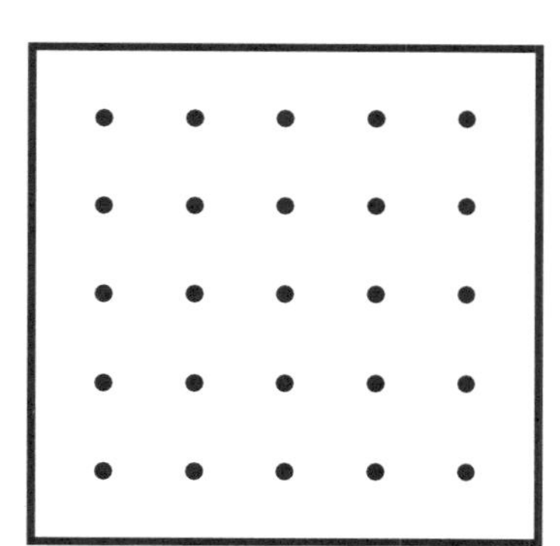

Circular Geoboard Recording Sheet

Name ....

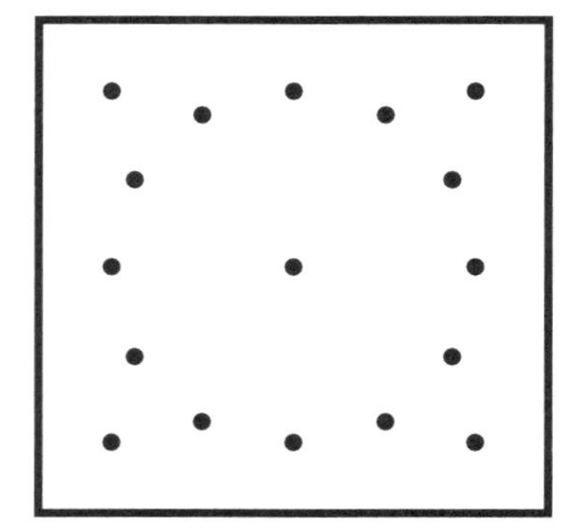
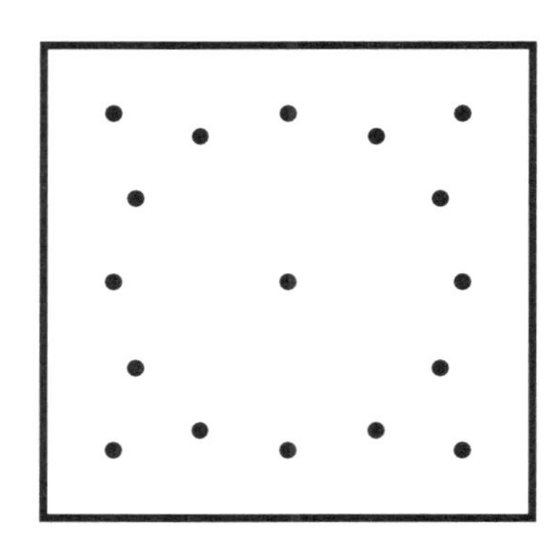
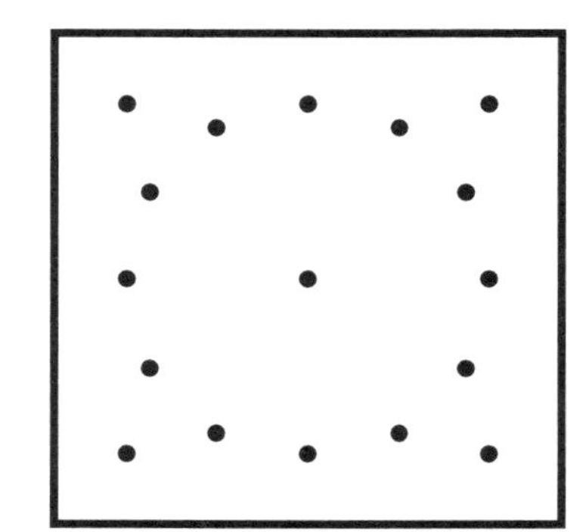
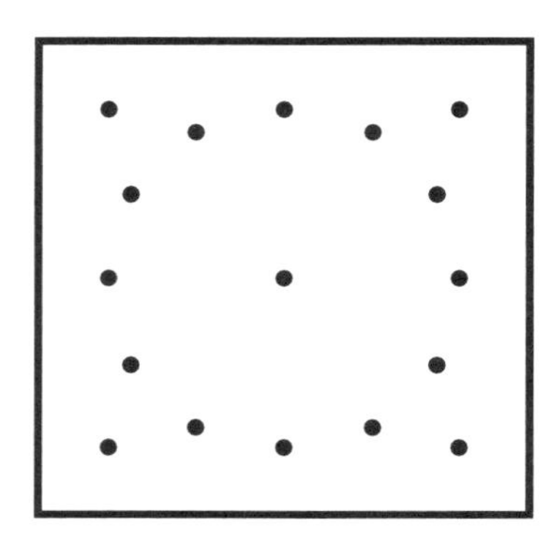
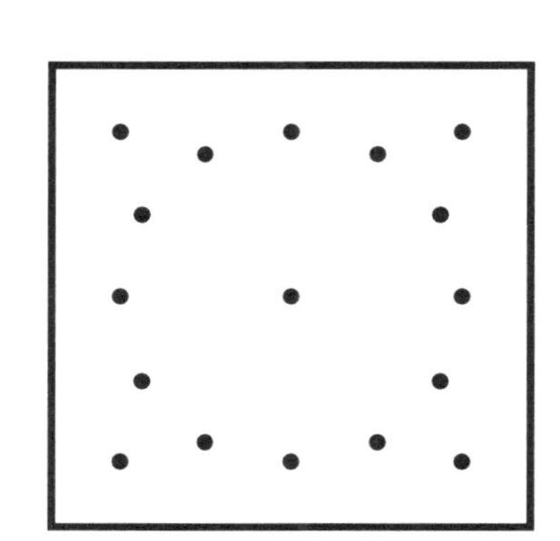
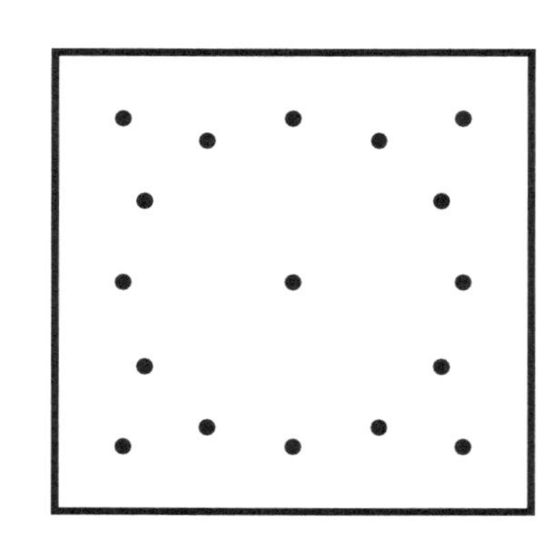
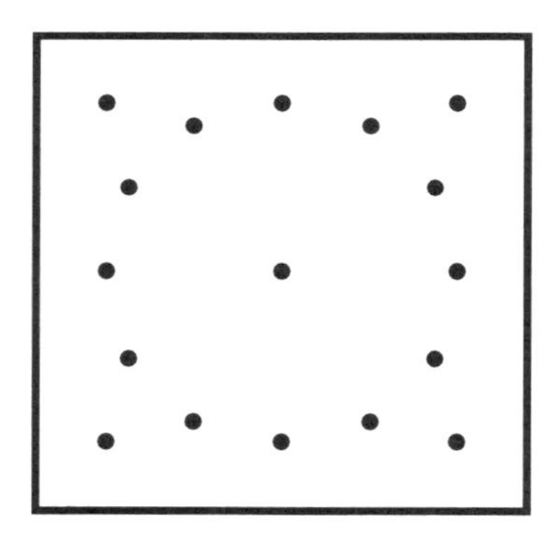
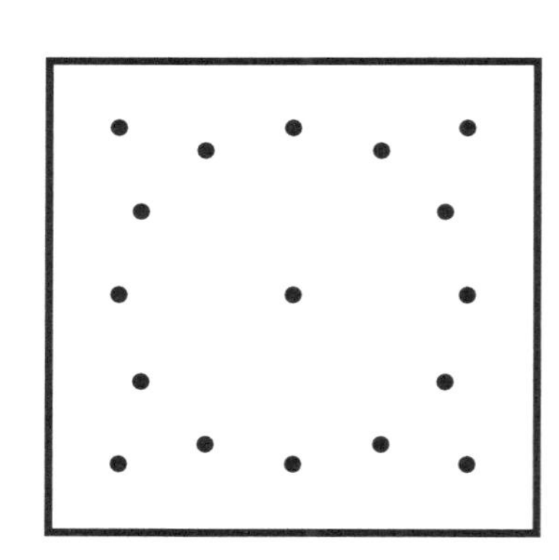
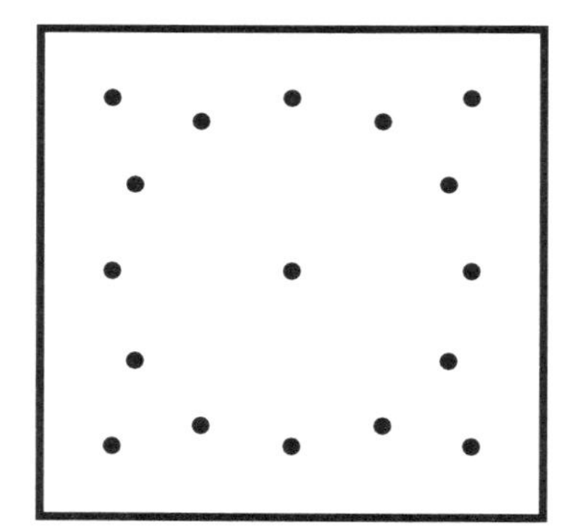